TEST BANK FOR

GEOMETRY
SEEING, DOING, UNDERSTANDING

THIRD EDITION

Harold R. Jacobs

Prepared By
James B. Dressler

W.H. FREEMAN AND COMPANY
New York

07167-5608-0
(c) 2004 by Harold R. Jacobs. All rights reserved.
Printed in the United States of America
First printing 2003

W. H. Freeman and Company
41 Madison Avenue
New York, NY 10010
Houndmills, Basingstoke RG21 6XS, England
www.whfreeman.com

Contents

INTRODUCTORY COMMENTS
CHAPTER TESTS
MIDTERM EXAMINATION
FINAL EXAMINATION
ANSWERS

Introductory Comments

This book consists of chapter tests, a midyear examination, and a final examination that may be used with *Geometry: Seeing, Doing, Understanding*, Third Edition. Two versions, identified by the letters A and B in the lower right hand corner of each page, are provided for each examination. Complete answers for all the tests are in a separate section at the end of this book.

I have found that most classes benefit from frequent short quizzes in addition to chapter tests; quizzes encourage regular study and discourage the temptation to cram at irregular intervals. I have not included any quizzes in this book, however, because I feel that they should be spontaneous and adapted to each class.

GEOMETRY
CHAPTER TESTS

GEOMETRY: Test on Chapter 1   Name_____

1) State whether each of the following are true or false.

   a) A *point* can be described as "that which has no part".

   b) A *line segment* is not bounded.

   c) A *line* and a *line segment* are the same thing.

   d) An *angle* is a pair of rays that have the same endpoint.

2) Complete these definitions.

   a) Points are *coplanar* if there is a plane that contains…

   b) Lines are *concurrent* if they contain …

   c) Points are *noncollinear* if no single line …

3) Sketch an example of three Collinear points A, B and C.

4) Use a protractor and the triangle below to measure:

   a) ∠A (include units)

   b) ∠B (include units)

   c) ∠C (include units)

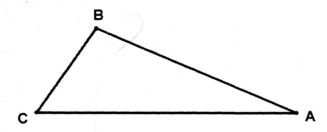

5) Determine if each of the following is a: *triangle, quadrilateral, pentagon, hexagon* or *octagon*.

   a)

   b)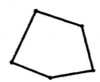

   c)

   d)

   e)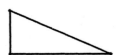

TURN OVER

GEOMETRY: Test on Chapter 1   Name_____

6) Given a square with the length of each side equal to 3 inches:

   a) What is the area? (include units)

   b) What is the perimeter? (include units)

   c) Is this square a rectangle?

7) Draw a 1 to 2 inch line using a straight edge. *Bisect* the line.

8) Draw an angle ABC, *bisect* the angle using a straight edge and compass.

9) If a 90° angle is bisected properly, what will the measure of each of the two resulting angles be?

10) Are *polyhedra* one, two or three-dimensional?

END

GEOMETRY: Test on Chapter 1   Name_____

1) Use the figure below to:

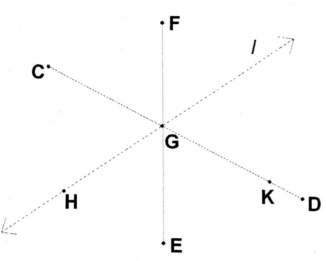

a) List any four *line segments*.

b) Identify a *line*.

c) Name all *collinear points* on line segment CD.

d) Name any three *noncollinear points*.

2) Perform the following using the figure below.

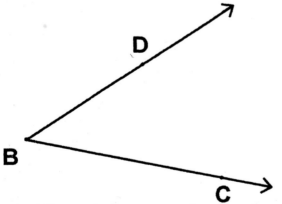

a) Name the two rays that are the sides of ∠B.

b) Use a straight edge to draw line segment CD. What is the length of CD in centimeters?

c) Use a protractor to measure ∠B. What is that measure?

3) Indicate the number of *dimensions* for each figure:

a) polyhedron

b) line segment

c) polygon

4) Use a ruler and protractor to draw a square whose sides are 2 inches long.

a) What is the area of the square?

b) What is the perimeter of the square?

c) Is this square a rectangle?

5) Determine if each of the following is a *quadrilateral, pentagon,* or *octagon.*

a)

b)

c)

TURN OVER

1B-1

GEOMETRY: Test on Chapter 1   Name_____

6) Using a ruler and compass, perform the following construction:

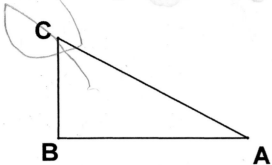

   a) Bisect angle A.

   b) Bisect side AB.

7) Use a compass to draw a circle whose diameter is 4 inches.

8) How many sides does a:

   a) hexagon have?

   b) triangle have?

9) If a 90° angle is bisected properly and the resulting angles are again bisected, what will the measure of each of the resulting angles be?

10) Name three concurrent lines in the figure.

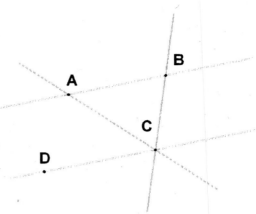

END

GEOMETRY: Test on Chapter 2   Name_____

1. Consider the statement, "The clouds are gray, if rain is falling."

   a) What is the hypothesis?

   b) What is the conclusion?

   c) In symbolic form, a → b, what is a and what is b?

2. Give the "if-then" statement suggested by each of the following.

   a) When the new baby girl is smiling, she is happy.

   b) Clean your room so that you can go to the concert.

   c) To win the game your score must be the highest.

   d) When the Giant Panda is hungry, it searches for bamboo.

3. Consider the definition: A horizontal line is a line that is perpendicular to a vertical line.

   a) Write the above definition in "if-then" form.

   b) Write the converse.

   c) In this case, is the converse true? Why or why not?

4. Read the following and mark each true or false.

   a) An Euler diagram may be used to represent the conditional statement "If a, then b."

   b) In an indirect proof, the opposite of the conclusion is supposed. When a contradiction is found, the opposite of the conclusion is proven to be false, so the conclusion is true.

   c) The converse of a → b is a → a.

   d) A statement and its converse must have the same meaning.

5. Consider the following argument:

   If you live in New York City, then you live in New York State.
   If you live in New York State, then you live in the United States.
   If you live in the United States, then you live in North American.
   Therefore, if you live in New York City, then you live in North America.

   a) List the premises of the above argument?

   b) What is the conclusion?

   c) Is the argument a *syllogism*?

TURN OVER

GEOMETRY: Test on Chapter 2   Name_____

6. Draw an Euler diagram to represent the following statement:

   If a discovery is accidental, then it is serendipitous.

7. What is the missing statement in the syllogism?

   (missing statement)

   She would run the fastest, if she would join the track team.

   If she would run the fastest, she would win all of the races.

   Therefore, if she enjoyed running, she would win all of the races.

8. Identify each of the following as a **Definition** or a **Postulate**.

   a) Points are collinear iff there is a line that contains all of them.

   b) Lines are concurrent iff they contain the same point.

   c) Three noncollinear points determine a plane.

   d) Points are coplanar iff there is a plane that contains all of them.

   e) Two points determine a line.

9. State the beginning assumption to prove the following theorem using an **indirect** proof.

   Theorem. The sum of two even numbers is an even number.

   Proof.

   (What is the beginning assumption?)

10. Use the figure to find the following:

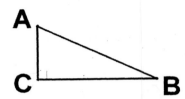

   a) ∠A + ∠B

   b) ∠B, if ∠A = 30°

   END

2A-2

GEOMETRY: Test on Chapter 2   Name_____

1. Consider the statement, "You can have dessert, if you eat all of your vegetables."

   a) What is the hypothesis?

   b) What is the conclusion?

   c) In symbolic form, a → b, what is a and what is b?

2. Give the "if-then" statement suggested by each of the following.

   a) When there is a line that contains all of them, points are collinear.

   b) To get good grades, you must study hard.

   c) To win the game, your score must be the highest.

3. Consider the definition: Circles are *coplanar*, iff they lie in the same plane.

   a) Write an "if-then" statement suggested by the above definition.

   b) Write the converse.

   c) In this case, is the converse true? Why or why not?

4. Read the following and mark each true or false.

   a) An Euler diagram may be used to represent the conditional statement "If a, then b."

   b) A *syllogism* is an argument in the form a → b, a → c, therefore b → c.

   c) The *converse* of a → b is b → a.

   d) A *theorem* is just a definition.

5. Consider the following argument:

   If this is a direct proof, then this is not an indirect proof.
   If this is not an indirect proof, then there is no contradiction.
   Therefore, if this is a direct proof, then there is no contradiction.

   a) What are the premises of the above argument?

   b) What is the conclusion?

   c) Is the argument a *syllogism*?

6. Draw an Euler diagram to represent an argument in the form:

   a → b
   b → c
   Therefore, a → c.

7. What is a *postulate?*

8. Identify each of the following as a **Definition** or **Postulate**.

   a) Two points determine a line.

   b) Points are collinear iff there is a line that contains all of them.

   c) Points are coplanar iff there is a plane that contains all of them.

TURN OVER

2B-1

GEOMETRY: Test on Chapter 2   Name_____

9. Write the "if-then" statement represented by the following Euler diagram.

10. Use the figure below to answer the following. The radius of c_1 = 10 cm.

a) If the radius of c_1 is twice the radius of c_2, then what is the diameter of c_2?

b) If the radius of c_1 is twice the radius of c_2, then what is the exact circumference of c_2?

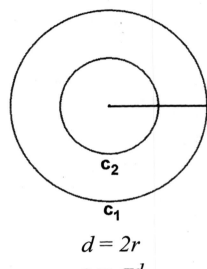

$d = 2r$
$c = \pi d$

END

GEOMETRY: Test on Chapter 3   Name_____

1) Identify each of the following algebraic properties of equality.

   a) If $a = b$, then $a + c = b + c$.

   b) If $a = b$, then $ac = bc$.

   c) $a = a$.

2) Indicate which property of equality is illustrated by each of the following statements.

   a) If $x + y = x + z$, then $y = z$.

   b) If $d = 2r$, then $\frac{d}{2} = r$.

   c) If $a = b$ and $a + c = 10$, then $b + c = 10$.

   d) If $\angle A + \angle B + \angle C - 90° = 90°$, then $\angle A + \angle B + \angle C = 180°$.

3) Complete the Betweenness of Points Theorem.

   If A-B-C, then AB + ...Complete.

4) Complete the Definition of Betweenness of Points.

   A-B-C iff $a < b < c$ or ...Complete.

5) With X = 5, Y = 10 and Z = 15, indicate whether each of the following are True or False.

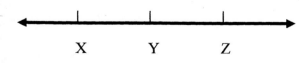

   a) Z-Y-X, because 15 > 10 > 5.

   b) XY + YZ = XZ, since X-Y-Z.

   c) XZ + XY = XZ + YZ, since XY = YZ.

   d) X-Z-Y, because 5 < 10 < 15.

   e) X-Y-Z, because 5 < 10 < 15.

6) Answer the following based on the figure below.

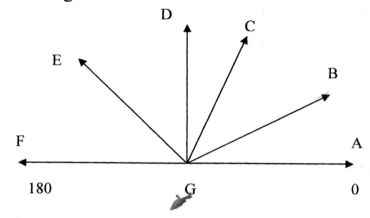

   a) Is $\angle AGE$ acute?

   b) Are $\angle AGD$ and $\angle FGD$ Supplementary?

   c) Is ray GB between ray GA and ray GF?

   d) What is the measure of $\angle BGE$, if $\angle AGB = 40°$ and $\angle AGE = 135°$?

   e) Write the angular equation that follows from the fact that GC-GD-GE.

   f) If ray GE bisects $\angle DGF$, does $\angle DGE = \angle EGF$?

   g) Is $\angle AGF$ a straight angle?

TURN OVER

3A-1

GEOMETRY: Test on Chapter 3   Name_____

7) If ∠1 is complementary to ∠2, does ∠1 + ∠2 + 90° = 180°?

8) ∠B = 40° and ∠C = 50°, answer the following.

   a) Are ∠B and ∠C complementary?

   b) If ∠E is complementary to ∠C why does ∠B = ∠E?

9) Refer to the figure below to answer the following.

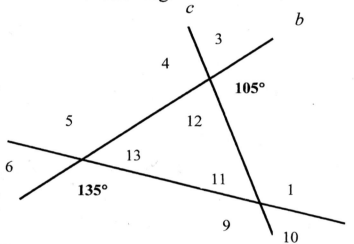

   a) Why does ∠3 = ∠12?

   b) How many degrees does ∠12 equal?

   c) How many degrees does ∠11 equal?

   d) Is line *b* perpendicular to line *c*?

   e) Name two angles equal to 75°.

   f) Are ∠1 and ∠11 a linear pair?

10) Answer the following.

   a) Can a 90° angle be formed by two non-perpendicular lines?

   b) If two lines form a right angle, are the lines perpendicular?

   c) If two lines do not lie in the same plane, can they be parallel to one another?

END

3A-2

GEOMETRY: Test on Chapter 3   Name_____

1) Identify each of the following algebraic properties of equality.

   a) If $a = b$, then $a - c = b - c$.

   b) If $a = b$ and $c \neq 0$, then $\dfrac{a}{c} = \dfrac{b}{c}$.

   c) $a = a$.

2) Indicate which property of equality is illustrated by each of the following statements.

   a) If $d = 2r$, $5d = 10r$.

   b) If $(2x + 1) + y = (2x + 1) + z$, then $y = z$.

   c) If $\angle A - 100° = 90°$, then $\angle A = 190°$.

   d) If $a = b$ and $a + 5 = 10$, then $b + 5 = 10$.

3) Complete the Definition of Betweenness of Points.

   A-B-C iff $a < b < c$ or …*Complete*.

4) Complete the Betweenness of Points Theorem.

   If A-B-C, then AB + BC…*Complete*.

5) With W = 0, X = 7, Y = 14 and Z = 21, indicate whether each of the following are True or False.

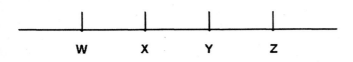

   a) XZ = 21 − 7.
   b) WZ = XZ.

   c) WX + XZ = WZ, since W-X-Z.
   d) X-Z-Y, because 7 < 14 < 21.
   e) W-X-Y, because 0 < 7 < 14.

6) Answer the following based on the figure below.

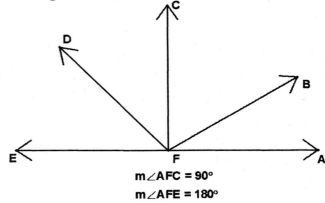

m∠AFC = 90°
m∠AFE = 180°

   a) Is ∠AFB acute?

   b) Are ∠AFC and ∠AFE Supplementary?

   c) Is ray FD between ray FA and ray FE?

   d) What is the measure of ∠CFB, if ∠DFB = 105° and ∠DFC = 45°?

   e) Which angle is a *straight* angle?

   f) If FD bisects ∠CFE, does ∠EFD = ∠CFD?

   g) Is there more than one *right* angle? If so, name them.

TURN OVER

3B-1

GEOMETRY: Test on Chapter 3   Name_____

7) If ∠1 is complementary to ∠2, what does ∠1 + ∠2 − 90° equal?

8) Given ∠X = 30° and ∠Y = 60°, answer the following.

   a) Why are ∠X and ∠Y not a linear pair?

   b) Why are ∠X and ∠Y not vertical angles?

9) Refer to the figure below to answer the following.

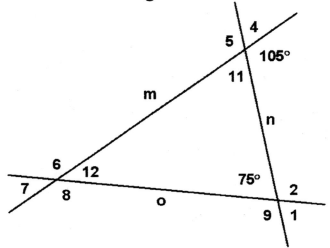

   a) Why does ∠7 = ∠12?

   b) How many degrees does ∠2 equal?

   c) How many degrees does ∠4 equal?

   d) Why are ∠1 and ∠2 supplementary?

   e) Name two angles equal to 105°.

   f) Why are line *m* and *n* not parallel?

10) Answer the following.

    a) Can two non-perpendicular lines form a 90° angle?

    b) If ∠A and ∠B are supplementary angles, when can ∠A = ∠B?

    c) If two lines do not lie in the same plane, can they be parallel to one another?

END

GEOMETRY: Test on Chapter 4   Name_____

1) State the distance formula for the distance between the points $P_1(x_1, y_1)$ and $P_2(x_2, y_2)$.

2) Answer the following based on the coordinate system.

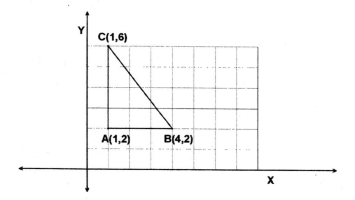

   a) Use the distance formula to calculate the length of AB.
   b) Use the distance formula to calculate the length of AC.
   c) Use the distance formula to calculate the length of BC.

3) Using the grid below, identify the coordinates of:

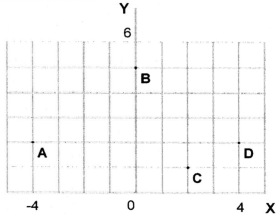

   a) Point A
   b) Point B
   c) Point C
   d) Point D

4) Yes or no, is the following triangle…

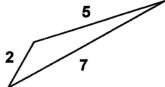

   a) Scalene?
   b) Acute?
   c) Equiangular?
   d) Obtuse?

5) Yes or no, is the following triangle…

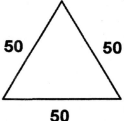

   a) Equilateral?
   b) Acute?
   c) Equiangular?
   d) Isosceles?

6) **Given**: equilateral $\triangle ABC$ and the fact that BD bisects $\angle ABC$.
   Prove: $\triangle ABD \cong \triangle CBD$ (using SAS)

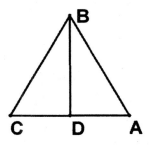

TURN OVER   4A-1

GEOMETRY: Test on Chapter 4   Name_____

7) Answer the following either True or False.

   a) A right triangle must have two 90° angles.

   b) Two perpendicular lines serve as axes in a two-dimensional coordinates system.

   c) If two angles of a triangle are equal, the sides opposite them are equal.

8) Given: ∠B = 121°, ∠C = 23°
   ∠E = 23°, ∠G = 121°
   ∠H = 23° and the figure below, answer the following.

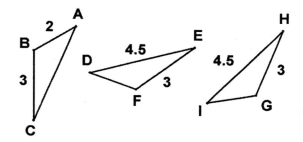

   a) What postulate indicates that △ABC ≅ △IGH?

   b) What postulate indicates that △DFE ≅ △IGH?

   c) Why is △DFE ≅ △ABC?

9) Given ∠B = ∠C = ∠H, use the right triangles below to answer the following.

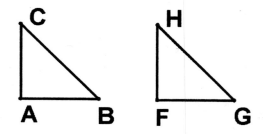

   a) Why does AB = AC?

   b) What line segment does FG equal?

10) Use a straight edge and compass to copy and Bisect the following angle.

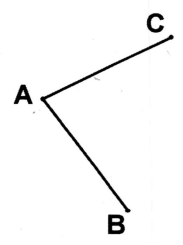

END

GEOMETRY: Test on Chapter 4   Name_____

1) Answer the following either True or False.

   a) Choosing an origin and ruler, gives us a *two-dimensional coordinate system.*

   b) Two perpendicular lines called *axes* allow us to locate points in a plane.

   c) The axes in a two-dimensional coordinate system form four quadrants.

2) Fill in the blanks to complete the following definition of congruent.

   Two triangles are *congruent* iff there is a correspondence between their vertices such that all of their corresponding _____ and _____ are equal.

3) Given: ∠A = 35°, ∠B = 122°, ∠C = 23°, ∠F = 122°, ∠G = 122°, ∠H = 23° and the figure below, answer the following.

   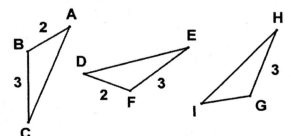

   a) What postulate indicates that △ABC ≅ △DFE?

   b) What postulate indicates that △ABC ≅ △IGH?

   c) Why is △DFE ≅ △IGH?

4) Fill in the blank to complete the theorems.

   a) If two sides of a triangle are equal, the angles opposite them are _____.

   b) If three sides of one triangle are equal to three sides of another triangle, the triangles are _____.

5) **Given**: square ABCD and the fact that AC bisects ∠DAB and ∠BCD.
   Prove: △ABC ≅ △ADC (using ASA)

   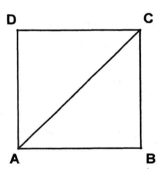

6) Answer the following True or False.

   A triangle is…

   a) *obtuse* iff it has an obtuse angle.

   b) *scalene* iff all of its angles are equal.

   c) *isosceles* iff it has at least two equal sides.

   d) *right* iff it has a right angle

   TURN OVER

GEOMETRY: Test on Chapter 4   Name_____

7) Using the figure below, give the reason for each of the following.

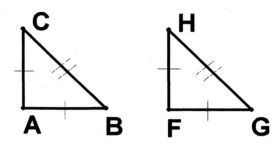

a) ∠B = ∠C.

b) △ABC ≅ △FGH

8) Given ∠1 = ∠2 = ∠3, answer the following

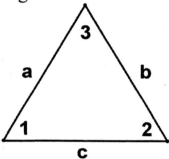

a) Does $a = b$?

b) Is the triangle equilateral?

c) What is the measure of ∠2?

9) Use a straight edge and compass to construct a copy of the following triangle.

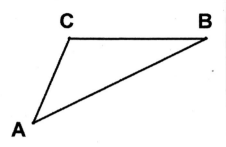

10) Answer the following based on the figure.

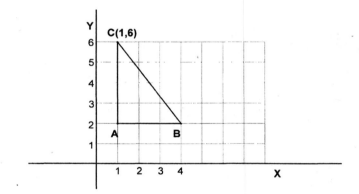

a) What are point A's coordinates?

b) What are point B's coordinates?

c) What is the distance between points A and B.

d) What is the distance between points A and C.

e) What is the distance between points B and C.

END

4B-2

GEOMETRY: Test on Chapter 5   Name_____

1) Name the property of equality or inequality illustrated by each of the following statements.

   a) Either $e < 2.72$ or $e = 2.72$ or $e > 2.72$.

   b) Because $e < \dfrac{27}{10}$, $10e < 27$.

   c) If $\angle A < \angle B$, then $\angle A + 90° < \angle B + 90°$.

2) Complete the Addition Theorem of Inequality.

   If $a > b$ and $c > d$, then … *complete*.

3) Complete the "Whole Greater than Part" Theorem.

   If $a > 0$, $b > 0$, and $a + b = c$, then … *complete*.

4) Use the figure below to answer the following.

   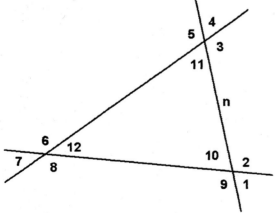

   a) Which angles are exterior angles of the triangle?

   b) If $\angle 1 + \angle 7 = 115°$, what does $\angle 4$ equal?

   c) If $\angle 6 = 120°$, can $\angle 11 = 121°$?

5) Answer the following True or False, based on the figure below.

   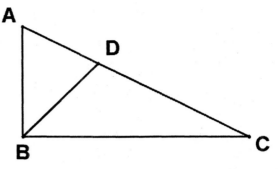

   a) If $\angle A > \angle C$, then $AB > BC$.

   b) $\angle A < \angle ADB$, so $BD < AB$.

   c) $BC < AC$, so $\angle A < \angle ABC$.

   d) $DC > BD$ and $BC > DC$, so $\angle BDC$ is the largest angle of $\triangle BCD$.

   e) $AB + BC = AC$.

6) Complete Theorem 13.

   If two sides of a triangle are unequal, then … *complete*.

7) Given $\triangle XYZ$ (not shown here), with $XY = 3$, $YZ = 4$ and $ZX = 5$. Is $\angle X$, $\angle Y$ or $\angle Z$ the largest angle?

8) Answer the following correctly using either Yes or No.

   a) Can an exterior angle of a triangle be equal to either of the remote interior angles of the triangle?

   b) Is the sum of any two sides of a triangle greater than the third side.

TURN OVER

GEOMETRY: Test on Chapter 5   Name_____

9) Fill in the blank using information from the following figure.

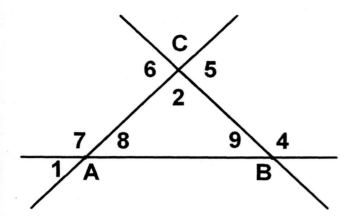

a) AB + BC > _____.

b) ∠4 + ∠9 = _____ degrees.

c) ∠2 + ∠8 + ∠9 = _____ degrees.

d) ∠4 = ∠_____ + ∠_____.

e) ∠7 > ∠_____ and ∠7 > ∠_____.

10) Provide reasons for the steps in the following proof.

Given: △ABC and with ∠BAC < ∠2
Prove: BD < AD

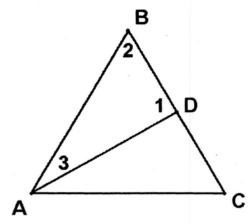

Proof.

a) ∠BAC < ∠2.

b) ∠3 < ∠BAC.

c) ∠3 < ∠2.

d) BD < AD

5A-2

GEOMETRY: Test on Chapter 5   Name_____

1) Name the property of equality or inequality illustrated by each of the following statements.

   a) If $\angle C < 75° - \angle B$ and
      $\angle B < \angle C$, then
      $\angle B < 75° - \angle B$.

   b) If $\angle A < 45° - \angle A$, then
      $2\angle A < 45°$

   c) Either $2\pi > 6.28$, $2\pi = 6.28$, or $2\pi < 6.28$.

2) Fill in the blank to complete the Addition Theorem of Inequality.

   If $a > b$ and _____ , then
   $a + c > b + d$.

3) Fill in the blank to complete the "Whole Greater than Part" Theorem.

   If $a > 0$, $b > 0$, and _____ ,
   then $c > a$ and $c > b$.

4) Use the figure below to answer the following.

   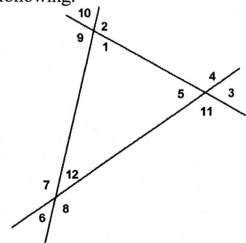

   a) Which angles are *not* exterior angles of the triangle?

   b) If $\angle 4 + 80° = 180°$, what does $\angle 4$ equal?

   c) If $\angle 6 = 45°$, can $\angle 11 = 119°$.

5) Answer the following True or False, based on the figure below.

   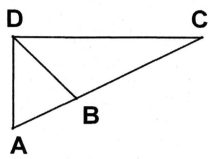

   a) If $\angle A > \angle C$, then $AD > DC$.

   b) $\angle A < \angle ADB$,
      therefore $BD < AB$.

   c) $BC < DC$, so $\angle CDB < \angle B$.

   d) $BC > BD$ and $CD > BC$, so $\angle DBC$ is the greatest angle of $\triangle BCD$.

   e) $DB + BC > DC$.

6) Complete Theorem 13.

   If two angles of a triangle are unequal, then ...*complete*.

7) Given $\triangle ABC$ (not shown here), such that $AB = 3.0$ cm, $BC = 3.15$ cm and $CA = 2.0$ cm. Which angle is the least: $\angle A$, $\angle B$ or $\angle C$?

TURN OVER

GEOMETRY: Test on Chapter 5   Name_____

8) Answer the following correctly using either Yes or No.

   a) Can an exterior angle of a triangle be equal to the sum of the remote interior angles of that triangle?

   b) Is the sum of any two sides of a triangle equal to the third side.

9) Fill in the blank using information from the following figure.

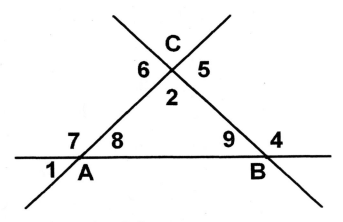

   a) AC + BC > _____.

   b) ∠2 + ∠6 = _____ degrees.

   c) ∠2 + ∠8 + ∠9 = _____ degrees.

   d) ∠5 = ∠_____ + ∠_____.

   e) ∠4 > ∠_____ and ∠4 > ∠_____.

10) Using the figure, provide reasons for the following proof.
    Given: Right △ABC
    Prove: AC > AB

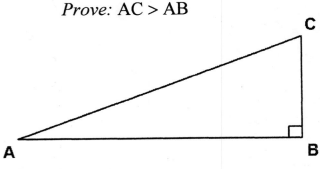

Proof.
   a) ∠B = 90°.
   b) ∠A + ∠B + ∠C = 180°.
   c) ∠A + 90° + ∠C = 180°.
   d) ∠A + ∠C = 90°.
   e) ∠A + ∠C = ∠B.
   f) ∠B > ∠C.
   g) AC > AB.

END

GEOMETRY: Test on Chapter 6   Name_____

1) Indicate whether each of the following are true or false.

   a) In a plane, a line perpendicular to one of two parallel lines is also perpendicular to the other.

   b) The acute angles of a right triangle are not complementary.

   c) Each angle of an equilateral triangle is 45°.

   d) The sum of the angles of a triangle is 180°.

2) Answer the following based on the parallel lines and transversal below.

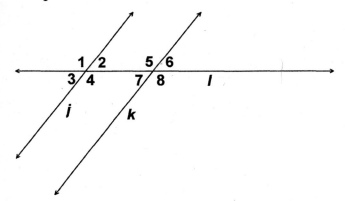

   a) Which line is the *transversal*?

   b) List all pairs of *corresponding angles*.

   c) List all pairs of *alternate interior angles*.

   d) List all pairs of *interior angles on the same side of the transversal*.

   e) Which lines are *parallel*?

3) Give the reasons for the following proof that equal alternate interior angles mean that lines are parallel.

   Given: ∠1 = ∠2
   Prove: l // m.

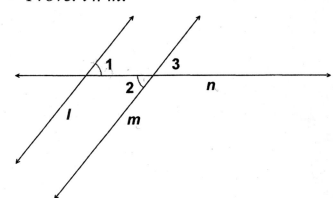

   Proof.
   a) ∠1 = ∠2.
   b) ∠2 = ∠3.
   c) ∠1 = ∠3.
   d) Line *l* is parallel to line *m*.

4) Given lines *j*, *k*, and *l* all in the same plane, answer the following.

   a) If *j* // *k* and *k* // *l*, is *j* // *l*?

   b) If *j* // *k* and *k* ⊥ *l*, is *j* ⊥ *l*?

   c) If *j* intersects *k*, is *j* // *k*?

   d) If *k* and *l* do not intersect, is *k* // *l*?

TURN OVER

GEOMETRY: Test on Chapter 6   Name_____

5) Given △ABC and j // AB, answer the following.

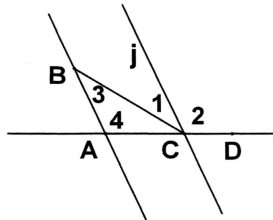

a) Why does ∠1 = ∠3?

b) Why does ∠2 = ∠4?

c) Why does ∠BCD = ∠1 + ∠2?

6) Use the figure to answer the following.

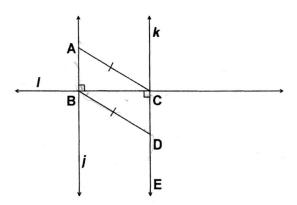

a) Why is △ABC ≅ △DCB?

b) Why is line j // k?

c) Why does ∠BDE = ∠DCB + ∠DBC?

7) Answer the following correctly using either Yes or No.

  a) Can an exterior angle of a triangle be less than either of the remote interior angles of the triangle?

  b) Can the sum of the angles of a triangle be less than 180°?

  c) Can four lines be parallel to each other?

  d) Can two lines be both parallel and perpendicular to each other?

8) What is the *Angle Sum Theorem*?

9) Use a straight edge and compass to construct a line perpendicular to *j*, through point A.

10) What does the *Parallel Postulate* state?

END

GEOMETRY: Test on Chapter 6   Name_____

1) Indicate whether each of the following are true or false.

   a) Two points are *symmetric with respect to a line* iff the line is the perpendicular bisector of the line segment joining the two points.

   b) A *transversal* is a line that intersects two or more lines in different points.

   c) In a plane, two points each equidistant from the endpoints of a line segment determine the perpendicular bisector of the line segment.

   d) Two lines are parallel iff they do not lie in the same plane and do not intersect.

2) Given that $j \parallel k$...

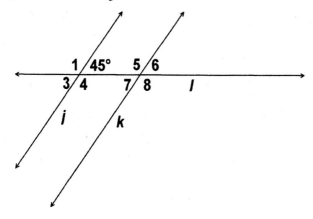

   a) What is the measure of ∠7?

   b) What is the measure of ∠6?

   c) What is the measure of ∠5?

   d) What is the measure of ∠1?

   e) Which line is the *transversal*?

3) Give the reasons for the following proof, that supplementary interior angles on the same side of the transversal mean that lines are parallel.

   Given: ∠1 and ∠2 are supplementary.
   Prove: $l \parallel m$.

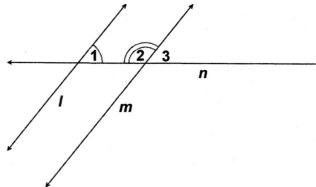

   Proof:

   a) ∠1 and ∠2 are supplementary.

   b) ∠2 and ∠3 are supplementary.

   c) ∠1 = ∠3.

   d) $l \parallel m$.

4) Given lines j, k, and l all in the same plane, answer the following.

   a) If $j \parallel k$ and $k \parallel l$, is $j \parallel l$?

   b) If j and k intersect, must j be perpendicular to l?

   c) If j intersects k, is $j \parallel k$?

   d) If j is not parallel to k and k is not parallel to l, must $j \parallel l$?

TURN OVER

GEOMETRY: Test on Chapter 6   Name_____

5) Complete the following Corollaries.

   a) If two angles of one triangle are equal to two angles of another triangle the third… *Complete*

   b) The acute angles of a right triangle are… *Complete*

   c) Each angle of an equilateral triangle is… *Complete*

6) Use the figure to answer the following.

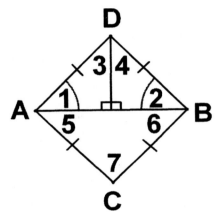

   a) Why does ∠3 = ∠4?

   b) If AB = AC, why does ∠5 = 60°?

   c) Why does ∠2 + ∠4 = 90°?

7) Using the figure, answer the following.

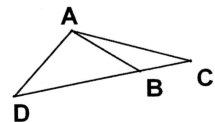

   a) Why does ∠ABC = ∠ADB + ∠DAB?

   b) Why can ∠BAC + ∠C + ∠ABC **not** equal 190°?

8) What is the *HL (hypotenuse leg) Theorem*?

9) Use the figure to answer the following and the fact AC = BD.

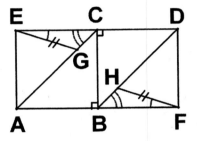

   a) Why is △ABC ≅ △DCB?

   b) Why is △BFH ≅ △CEG?

10) Use a straight edge and compass to construct a line parallel to *l*, through point P.

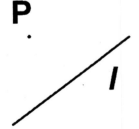

END

6B-2

GEOMETRY: Test on Chapter 7   Name_____

1) State whether each of the following are True or False.

   a) All triangles are convex.

   b) Quadrilaterals may be convex or concave.

   c) The sum of the angles of a quadrilateral is 180°.

   d) Each angle of a rectangle does not have to be 90°.

   e) A *parallelogram* is a quadrilateral whose opposite sides are parallel.

2) Answer the following based on the parallelogram below.

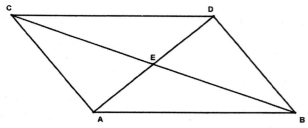

   a) If AB = 10, what length is CD?

   b) Why is AC parallel to BD?

   c) If AD and BC are diagonals and AE = 5, what length is ED?

   d) If ∠ACD + ∠CDB = 180° how many degrees is the sum of ∠DBA and ∠BAC?

3) Complete the given definitions.

   a) A *rhombus* is a quadrilateral all of whose sides are …

   b) A *square* is a quadrilateral all of whose sides and angles are…

   c) A *rectangle* is a quadrilateral all of whose angles are…

4) State whether the following are True or False.

   a) A square is also a rhombus.

   b) Any rectangle is also a square.

   c) A quadrilateral may be a square, rectangle or rhombus.

5) Given a quadrilateral ABCD, identify each as square, rectangle or rhombus.

   a) AB = BC = CD = DA

   b) AB = BC = CD = DA and ∠A = ∠B = ∠C = ∠D

   c) ∠A = ∠B = ∠C = ∠D

6) Explain in your own words what each of these are.

   a) A *trapezoid*.

   b) The *bases* of a trapezoid.

   c) An *isosceles trapezoid*.

7) Given triangle ABC with midsegment DE, answer the following.

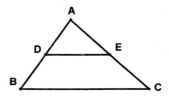

   a) If DE = 1.5 cm, what does BC equal?

   b) Are DE and BC parallel?

TURN OVER

7A-1

GEOMETRY: Test on Chapter 7   Name_____

8) Given the isosceles trapezoid ABDC with AC = BD...

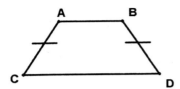

   a) Which sides are the legs?

   b) If ∠A = 123° what does ∠B equal?

   c) If ∠A = 120° what does ∠D equal?

   d) Are AB and CD parallel?

9) Given triangle ABC with midsegment DE, answer the following.

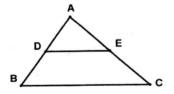

   a) If AD = 2 cm, what does DB equal?

   b) If EC = x, what does AC equal?

10) Complete the definition:

   A *midsegment* of a triangle is a line segment that joins the …

END

GEOMETRY: Test on Chapter 7   Name_____

1) State whether each of the following are True or False.

   a) All rectangles are *convex*.

   b) Two sides of a quadrilateral that intersect are called *opposite*.

   c) Each angle of a polygon is equal.

   d) Every quadrilateral has two diagonals.

2) Answer the following based on the parallelogram with diagonals BC and AD.

   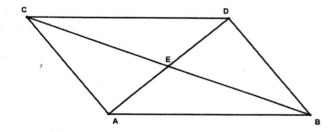

   a) If CE = 5 cm, what length is EB?

   b) If ∠ECD = 40°, what does ∠EBA equal?

   c) If ∠ACD + ∠CDB + ∠DBA = 280°, what does ∠BAC equal?

   d) What angle equals ∠CAB?

3) Fill in the blank to make the definition correct.

   a) A _____ is a quadrilateral all of whose sides and angles are equal.

   b) A _____ is a quadrilateral all of whose sides are equal.

   c) A _____ is a quadrilateral all of whose angles are equal.

4) State whether the following are True or False.

   a) The diagonals of a parallelogram bisect each other.

   b) All squares are also rectangles.

   c) All rectangles are parallelograms.

5) List the letters of the figures that are parallelograms. If a figure is not a parallelogram, tell why.

   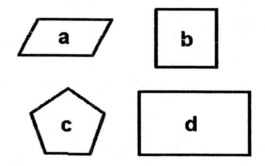

6) List the letters of the figures above, that have point symmetry.

7) Given triangle ABC with midsegment DE, answer the following.

   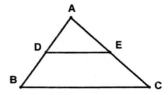

   a) If BC = 4 cm, what does DE equal?

   b) What kind of quadrilateral is DECB?

   TURN OVER

7B-1

GEOMETRY: Test on Chapter 7   Name_____

8) Given the isosceles trapezoid ABCD.

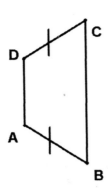

a) What makes this an *isosceles* trapezoid?

b) Does this isosceles trapezoid have point symmetry?

c) If ∠A = 110° what does ∠D equal?

d) If ∠B = 80° what does ∠D equal?

9) Given *rhombus* CDEF, answer the following.

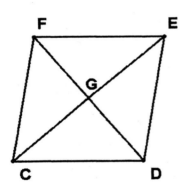

a) Is CDEF a parallelogram?

b) If FE = 2 ft, what does ED equal?

c) What is the measure of ∠G?

10) Indicate which figure below is a *concave* polygon.

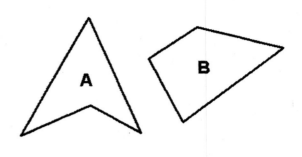

END

GEOMETRY: Test on Chapter 8   Name_____

1) State whether each of the following are True or False.

   a) An *isometry* is a transformation that preserves distance and angle measure.

   b) The fixed point about which a figure is rotated is called the *center* of rotation.

   c) A *dilation* does not reduce or enlarge an objects size.

   d) When performing a transformation on a figure, the result is called the *image*.

   e) A *dilation* is an example of an isometry.

2) Answer the following based on the figure below. FB is perpendicular to HD.

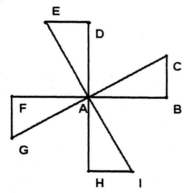

   a) Triangle ABC is rotated a magnitude of 180° about center A. Which triangle is its image?

   b) Line segment GA is reflected through EI. Which line segment is its image?

3) Complete the given definitions.

   a) A *rotation* is the composite of two successive reflections through … lines.

   b) A *translation* is the composite of two successive reflections through …lines.

   c) Two figures are *congruent* if there is an isometry such that one figure is the image of…

   d) A *glide reflection* is the composite of a translation and a reflection in a line parallel to the direction of the …

4) Identify the correct transformation: *translation* or *glide reflection*.

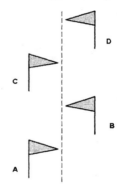

   a) Flag A to image C.
   b) Flag A to image B.
   c) Flag D to image A.
   d) Flag D to image B.
   e) Flag D to image C.

TURN OVER

8A-1

GEOMETRY: Test on Chapter 8   Name_____

5) Complete the given definitions.

   a) A figure has *reflection (line) symmetry* with respect to a line, if it coincides with its … …through the line.

   b) A figure has *rotation symmetry* with respect to a point, if it coincides with its …through less than 360° about the point.

   c) A pattern has *translation symmetry*, if it coincides with a …

6) The following is an equilateral hexagon inscribed in a circle.

   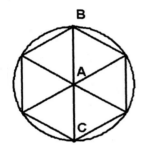

   a) What is the smallest angle that this image can be rotated to demonstrate *rotation symmetry?*

   b) Which point is the *center* of the symmetry?

7) Does the above figure have "*n*-fold" rotation symmetry? If so, what is value is *n*?

8) Answer the following based on the figure. Assume it extends endlessly in both directions.

   a) Does the pattern above have *translation symmetry?*

   b) Does the figure above have *reflection (line) symmetry?*

9) Does a circle have…

   a) translation symmetry?

   b) rotation symmetry?

10) Does a square have…

   a) rotation symmetry?

   b) translation symmetry?

END

GEOMETRY: Test on Chapter 8   Name_____

1) State whether each of the following are True or False.

   a) An *isometry* is a transformation that does not preserve distance and angle measure.

   b) Two figures are *congruent*, if there is an isometry such that one figure is the image of the other.

   c) A *dilation* does reduce or enlarge an objects size.

   d) A *reflection* is the result of two successive composite translations through parallel lines.

   e) A *rotation* is a glide reflection.

2) Answer the following based on the figure below.

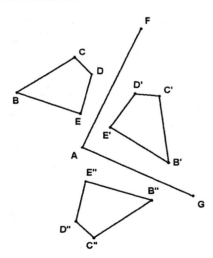

   a) Name the isometry which B'C'D'E' is the image of BCDE.

   b) Name the isometry which B''C''D''E'' is the image of BCDE.

3) Fill in the blank to complete the definition.

   a) A _____ is the composite of two successive reflections through intersecting lines.

   b) A _____ is the composite of a translation and a reflection in a line parallel to the direction of the translation.

   c) Two figures are _____ if there is an isometry such that one figure is the image of the other.

4) Identify the correct transformation: *translation, rotation, reflection* or *glide reflection*.

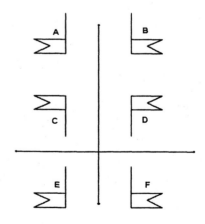

   a) Flag A to image E.
   b) Flag A to image B.
   c) Flag B to image E.
   d) Flag D to image E.
   e) Flag F to image C.

TURN OVER   8B-1

GEOMETRY: Test on Chapter 8   Name_____

5) Fill in the blank to complete the definitions.

   a) A figure has _____ symmetry with respect to a line, if it coincides with its reflection image through the line.

   b) A figure has _____ symmetry with respect to a point, if it coincides with its rotation image through less than 360° about the point.

   c) A pattern has _____ symmetry, if it coincides with a translation image.

6) The following is an equilateral triangle inscribed in a circle.

   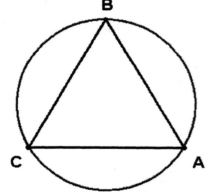

   a) What is the smallest angle that the figure may be rotated to demonstrate *rotation symmetry*?

   b) Where is the *center* of symmetry?

7) Does the above figure have "*n*-fold" rotation symmetry? If so, what value is *n*?

8) Answer the following based on the figure.

   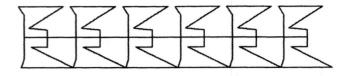

   a) Does the above pattern have *translation symmetry*?

   b) Does the above figure have *reflection (line) symmetry*?

9) Where is the *center* of the symmetry located for a rectangle?

10) Why does a square have *reflection (line) symmetry* about either one of its diagonals.

END

GEOMETRY: Test on Chapter 9   Name_____

1) State whether each of the following are True or False.

   a) The area of a right triangle is the product of its legs.

   b) The area of a triangle is half the product of any base and any corresponding altitude.

   c) The area of a square is the square of its sides.

   d) The area of a trapezoid is half the product of its altitude and the sum of its bases.

   e) The area of a rectangle is the product of its base and diagonal.

2) $\triangle ABF \cong \triangle DCE$. What is the area of trapezoid ADEF?

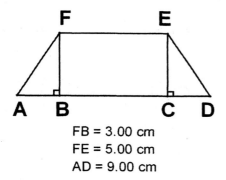

   FB = 3.00 cm
   FE = 5.00 cm
   AD = 9.00 cm

3) Fill in the blanks to make the equations true. HGFE is a parallelogram.

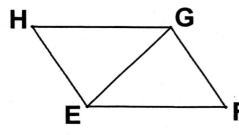

   a) $\alpha \triangle EFG = \alpha$_____.

   b) α_____ $= \alpha \triangle EFG + \alpha \triangle GHE$.

   c) α_____ $- \alpha \triangle GHE = \alpha \triangle EFG$.

4) Answer the following. The squares have area of 4 and 6 square units.

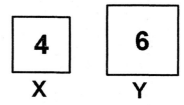

   a) What is the length of x?

   b) What is the **exact** length of y?

   c) What is x^2?

   d) What is y^2?

   e) What is $x^2 + y^2$?

   f) What is the decimal form of $\sqrt{x^2 + y^2}$?

   Give answer accurate to two decimal places.

5) Complete the formula for the square of a binomial.

   $$(a+b)^2 =$$

6) Answer the following using the square whose sides are each 2 feet.

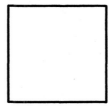

   2 feet

   a) What is the area in feet?

   b) What is the area in inches?

   TURN OVER

   9A-1

GEOMETRY: Test on Chapter 9   Name_____

7) Answer the following using the square whose sides are 36 inches.

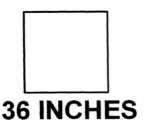

36 INCHES

   a) What is the perimeter in inches?

   b) What is the perimeter in feet?

8) What is the area of the following triangle?

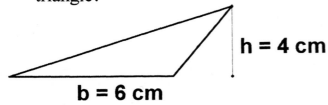

9) Use the figure to answer the following

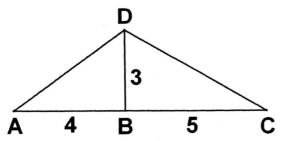

   a) What is the length of AD?

   b) What is CD^2?

   c) What is $AD^2 + CD^2$?

10) Give the reasons in the following proof.

Given: $\triangle ABC$ with base b and altitude h.

Prove: $\alpha \triangle ABC = \frac{1}{2} bh$

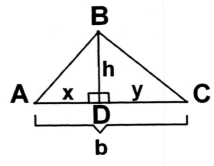

   a) $\alpha \triangle ABD = \frac{1}{2} xh$ and $\alpha \triangle CBD = \frac{1}{2} yh$.

   b) $\alpha \triangle ABC = \alpha \triangle ABD + \alpha \triangle CBD$.

   c) $\alpha \triangle ABC = \frac{1}{2} xh + \frac{1}{2} yh = \frac{1}{2} h(x + y)$.

   d) Because $b = x + y$, $\alpha \triangle ABC = \frac{1}{2} bh$.

END

GEOMETRY: Test on Chapter 9   Name_____

1) State whether each of the following are True or False.

   a) The area of a right triangle is half the product of its legs.

   b) The square of the hypotenuse of a right triangle is equal to the sum of the squares of the legs.

   c) If the square of one side of a triangle is not equal to the sum of the squares of the other two sides, the triangle is a right triangle.

   d) The area of a polygonal region is equal to the sum of the areas of its nonoverlapping parts.

   e) Congruent triangles do not have equal areas.

2) Use the parallelogram to answer the following.

   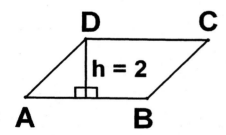

   a) If AB = 6 cm, what is the area of ABCD?

   b) If the other leg of the right triangle is also 2 cm, what is AD^2?

   c) What is the measure of the altitude of ABCD?

3) Fill in the blanks to make the equations true.

   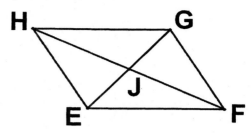

   a) $\alpha \Delta EGH = \alpha \Delta EJH + \alpha \Delta$_____.

   b) $\alpha EFGH = \alpha \Delta EGF + \alpha \Delta$_____.

4) Answer the following using the figure below. Assume that G, F, E, D and C are right angles.

   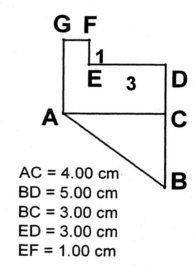

   AC = 4.00 cm
   BD = 5.00 cm
   BC = 3.00 cm
   ED = 3.00 cm
   EF = 1.00 cm

   a) What is the length of CD?

   b) What is the length of GF?

   c) What is the area of ΔABC?

   d) What is the area of the polygonal region ABDEFG?

5) Complete the formula for the square of a binomial.

   $$(a+b)^2 =$$

TURN OVER

GEOMETRY: Test on Chapter 9   Name_____

6) Answer the following using the figure.

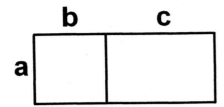

   a) What is the expression for just the area of the square?

   b) What is the expression for the total area?

   c) If a = b, b = 2 meters and c = 3 meters, what is the total area?

7) Answer the following using the triangle.

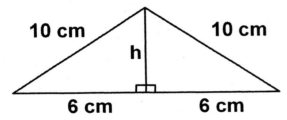

   a) What is the length of h?

   b) What is the entire area?

8) Which of the following has the greatest area?

   a) A right triangle with **leg** of 6 inches and a **hypotenuse** of 10 inches.

   b) A square whose **sides** are 4 inches.

   c) A trapezoid whose **altitude** is 10 inches and whose **bases** are 2 inches and 3 inches.

9) Complete the corollary.

   Triangles with equal bases and equal altitudes have equal…

10) Give the reasons in the following proof.

   Given: △ABC with base b and altitude h.

   Prove: $\alpha\triangle ABC = \frac{1}{2}bh$

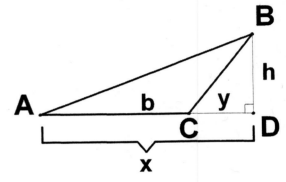

   a) $\alpha\triangle ABD = \frac{1}{2}xh$ and $\alpha\triangle CBD = \frac{1}{2}yh$.

   b) $\alpha\triangle ABD = \alpha\triangle ABC + \alpha\triangle CBD$.

   c) $\alpha\triangle ABC = \alpha\triangle ABD - \alpha\triangle CBD$.

   d) $\alpha\triangle ABC = \frac{1}{2}xh + \frac{1}{2}yh = \frac{1}{2}h(x-y)$.

   e) Because $b = x - y$, $\alpha\triangle ABC = \frac{1}{2}bh$.

END

GEOMETRY: Test on Chapter 10   Name_____

1) State whether each of the following are True or False.

   a) Two triangles similar to a third triangle are similar to each other.

   b) The ratio of the areas of two similar polygons is equal to the square of the ratio of the corresponding sides.

   c) A *proportion* is an inequality between two ratios.

   d) Corresponding altitudes of similar triangles have the same ratio as the corresponding sides.

   e) If two angles of one triangle are equal to two angles of another triangle, the triangles are similar.

2) **Given:** $\dfrac{a}{b} = \dfrac{c}{d}$

   Prove: $ad = bc$

3) Answer the following, given the proportion.

   $$\dfrac{a}{b} = \dfrac{c}{d}$$

   a) What are the *terms*?

   b) What are the *means*?

   c) What are the *extremes*?

4) Answer the following using the figure and the fact that DE // BC, AD = 15, DB = 45 and AE = 12.

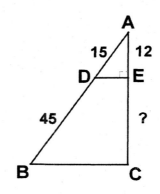

   a) Does $\dfrac{AD}{AB} = \dfrac{AE}{AC}$?

   b) Does $\dfrac{AD}{DB} = \dfrac{AC}{EC}$?

   c) What is the length of EC?

   d) What is the length of AC?

5) **Given:** In △ABC, AB = AC and DE // BC.

   Prove: AD = AE

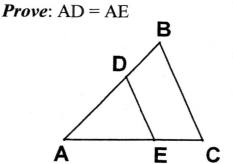

6) Why is △ABC similar to △EBD?

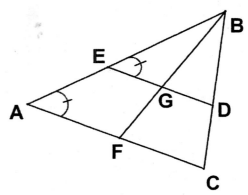

TURN OVER   10A-1

GEOMETRY: Test on Chapter 10   Name_____

7) Answer the following using similar polygons A and B and the fact that x > y.

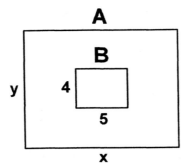

   a) What is αB?
   b) Given x = 15, what is αA?
   c) Given x = 20, what is y?

8) Suppose △ABC ~ △DEF and △GHI ~ △DEF. Why is △ABC ~ △GHI?

9) Answer the following using the similar polygons shown.

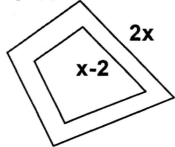

   a) If the ratio of the perimeters of the outer polygon to the inner polygon is $\frac{9}{3}$, why does $\frac{9}{3} = \frac{2x}{x-2}$?
   b) What is x?
   c) What is 2x?
   d) What is x − 2?

10) Are these two triangles similiar? Explain.

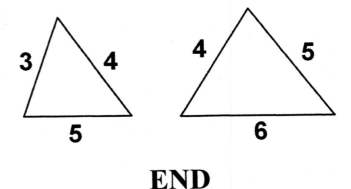

END

10A-2

GEOMETRY: Test on Chapter 10   Name_____

1) State whether each of the following are True or False.

   a) If a line parallel to one side of a triangle intersects the other two sides in different points, it cuts off segments not proportional to the sides.

   b) The *ratio* of the number a to the number b is the number $\frac{a}{b}$ (where $b \neq 0$).

   c) A *proportion* is an equality between two ratios.

   d) Two triangles similar to a third triangle are not similar to each another.

   e) If two angles of one triangle are equal to two angles of another triangle, the triangles not are similar.

2) Answer the following based on the triangle.

   a) What proportion is true for this triangle?

   b) What is x?

3) Answer the following based on the triangle.

   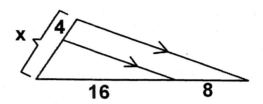

   a) What proportion is true for this triangle?

   b) What is x?

4) Why is △ABC similar to △A'B'C'?

   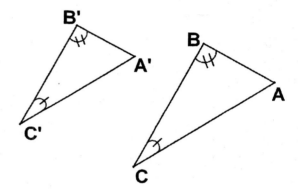

5) If △ABC ~ △A'B'C' and △A'B'C' ~ △A"B"C", why is △ABC ~ △A"B"C"?

   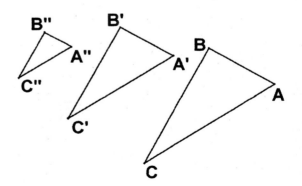

TURN OVER

10B-1

GEOMETRY: Test on Chapter 10   Name_____

6) Use the similar triangles and the fact that BG and EH are corresponding altitudes to answer the following.

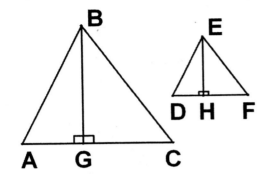

a) What is the ratio of the altitude of △ABC to the altitude of △DEF?

b) What is the proportion of the altitudes to AB and its corresponding side?

c) If $\dfrac{BG}{EH} = \dfrac{3}{2}$ and AB = 6, what does ED equal?

d) If $\dfrac{BG}{EH} = \dfrac{12}{3}$ and EF = 5, what does BC equal?

7) Answer the following using the similar polygons shown.

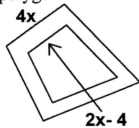

a) If the ratio of the perimeters of the inner polygon to the outer polygon is $\dfrac{3}{9}$, why does $\dfrac{3}{9} = \dfrac{2x-4}{4x}$?

b) What is x?

c) What is 4x?

d) What is 2x – 4?

8) If the area of polygon A is 36 square meters and the area of polygon B is 16 square meters and one of A's sides is 3 meters, what is the length B's corresponding side?

9) Give the reason for each step on the following proof.

GIVEN: In △ABC, AD bisects ∠BAC; AE = ED.

PROVE: $\dfrac{AE}{EC} = \dfrac{BD}{DC}$

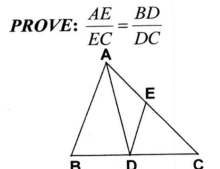

a) In △ABC, AD bisects ∠BAC.

b) ∠BAD = ∠DAC

c) AE = ED.

d) ∠EDA = ∠DAC.

e) ∠BAD = ∠EDA

f) ED // AB.

g) $\dfrac{AE}{EC} = \dfrac{BD}{DC}$.

10) What is the geometric mean of 3 and 12?

END

GEOMETRY: Test on Chapter 11   Name_____

1) State whether each of the following are True or False.

   a) In an isosceles right triangle, the hypotenuse is 2 times the length of a leg.

   b) Two non-vertical lines are parallel iff their slopes are equal.

   c) The altitude to the hypotenuse of a right triangle forms two triangles similar to it and to each other.

   d) Two non-vertical lines are perpendicular iff the product of their slopes is -1.

   e) In a right triangle, the hypotenuse is equal to the sum of the squares of the legs.

2) Use the right triangle ABC, to answer the following.

   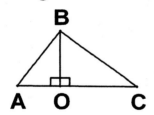

   a) Which segment is the hypotenuse?

   b) Which segment is the altitude to the hypotenuse?

   c) Which triangles are similar?

   d) What is $\dfrac{AO}{BO}$ proportionate to?

3) Use the Pythagorean Theorem to prove that in an isosceles right triangle, the hypotenuse is $a\sqrt{2}$.

Given: An isosceles right triangle with legs a and hypotenuse c.

Prove: $c = a\sqrt{2}$

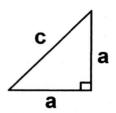

4) Answer the following using the 30° - 60° right triangle.

   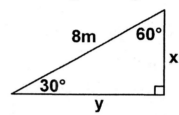

   a) How many square meters is $x^2 + y^2$?

   b) How many meters is x?

   c) How many meters is y?

5) Fill in the blank for the following definitions.

   a) The *tangent* of an acute angle of a right triangle is the ratio of the length of the _____ leg to length of the _____ leg.

   b) The *sine* of an acute angle of a right triangle is the ratio of the length of the _____ leg to length of the _____.

   c) The *cosine* of an acute angle of a right triangle is the ratio of the length of the _____ leg to length of the _____.

TURN OVER

GEOMETRY: Test on Chapter 11   Name_____

6) Use your calculator to find each of the following to *three decimal places*.

   a) What is cos 5°?

   b) What is sin 44°

   c) What is cos 75°?

   d) What is tan 15°?

   e) What is sin 60°?

7) Given the following ratios, use your calculator to find the measure of each angle to *three decimal places*.

   a) What is A if, cos A = 0.977?

   b) What is A if, sin A = 0.0123?

   c) What is B if, tan B = 0.25?

   d) What is C if, tan C = 7.8716?

   e) What is D if, sin D = 0.70711?

8) Complete the definition of *slope*.

   $$m = \frac{rise}{run} =$$

9) Use the △ABC to answer the following to the nearest degree.

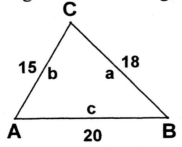

   a) What is the measure of ∠A?

   b) What is the measure of ∠B?

   c) What is the measure of ∠C?

10) Use the set of lines graphed below to answer the following.

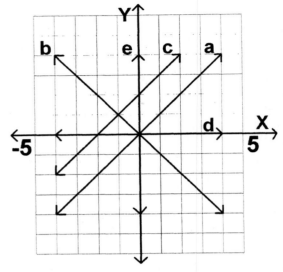

   a) What is slope of line a?

   b) What is slope of line b?

   c) Which line's slope is undefined?

   d) Which line has a slope of 0?

   e) Which lines are parallel?

   f) What line is perpendicular to c?

END

GEOMETRY: Test on Chapter 11   Name_____

1) State whether each of the following are True or False.

   a) Each diagonal of a square is $\sqrt{2}$ times the length of one side.

   b) The slope of a line is found by dividing the run by the rise.

   c) The cosine of an acute angle of a right triangle is the ratio of the length of the adjacent leg to the length of the hypotenuse.

   d) Two non-vertical lines are perpendicular iff the product of their slopes is 1.

   e) In a right triangle, the hypotenuse is equal to the sum of the legs.

2) Use the figure, to answer the following.

   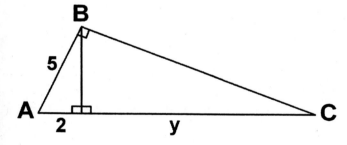

   a) What is the length of the projection of AB on AC?

   b) What is $\dfrac{2+y}{5}$ proportionate to?

   c) What is the length of y?

   d) What is the length of AC?

3) Use the square and its diagonal to answer the following.

   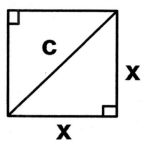

   a) What is length of C^2?

   b) What is the length of C?

   c) If $x = 10$ meters, what is the length of C?

4) Answer the following using the 30° - 60° right triangle.

   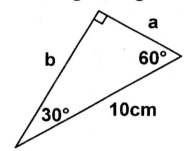

   a) How many square centimeters is $a^2 + b^2$?

   b) How many centimeters is a?

   c) How many centimeters is b?

5) In the following acute right triangle, when can cos A = sin A?

   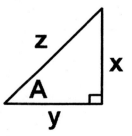

TURN OVER

11B-1

GEOMETRY: Test on Chapter 11   Name_____

6) Given the following information, copy the acute right triangle and label the legs and hypotenuse.

$$\tan A = \frac{2x}{y} \qquad \cos B = \frac{2x}{3z}$$

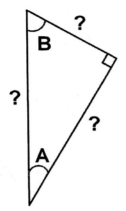

7) Answer the following using △ABC.

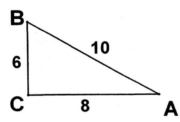

a) What is sin A as a fraction?

b) What is tan B as a fraction?

c) What is $\frac{\sin A}{\cos A}$ as a fraction?

d) What is ∠A to the nearest degree?

e) What is ∠B to the nearest degree?

8) If a street has a slope of 0.3, what is the street's angle of inclination to the nearest degree?

9) Use the △ABC to find the length of AC.

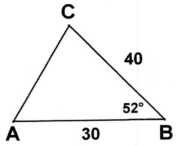

10) Use the set of lines graphed below to answer the following.

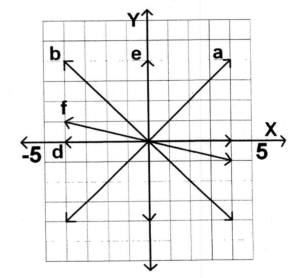

a) Which line has a slope of −1?

b) Which line has a slope of 1?

c) What is the slope of the line perpendicular to line e?

d) What is the slope of line e?

e) What is the slope of line f?

END

GEOMETRY: Test on Chapter 12   Name_____

1) State whether each of the following are True or False.

   a) If a line through the center of a circle is parallel to a chord, it also bisects it.

   b) The perpendicular bisector of a chord of a circle contains the center of the circle.

   c) In a circle, equal arcs have equal chords.

   d) An inscribed angle is equal in measure to half its intercepted arc.

   e) The tangent segments to a circle from an external point are not equal.

2) Give the reasons for the following proof.

   Given: OC ⊥ AB in circle O.

   Prove: OC bisects AB.

   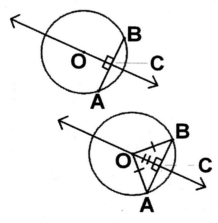

   a) Because OC ⊥ AB, ∠OCA and ∠OCB are right angles. Why?

   b) Draw OA and OB. Why?

   c) OA = OB. Why?

   d) OC = OC. Why?

   e) △ACO ≅ △BCO. Why?

   f) AC = BC. Why?

   g) OC bisects AB. Why?

3) Given that the chord JK is tangent to circle B, answer the following.

   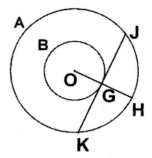

   a) Why can JK intersect circle B in only one point?

   b) If OG is a radius of circle B, how many degrees is ∠OGK?

   c) If OH is a radius of circle A, why does KG = GJ?

4) Complete the following.

   a) An inscribed angle is equal in measure to half its _____.

   b) A *secant* is a line that intersects a circle in _____.

   c) The tangent segments to a circle from an external point are _____.

   d) Inscribed angles that intercept the same arc are _____.

   TURN OVER

   12A-1

GEOMETRY: Test on Chapter 12   Name_____

5) Given △ABC is equilateral, answer the following.

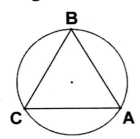

a) Which chords are equal?
b) Which arcs are equal?
c) Is ∠A an *inscribed* angle?
d) What is the measure of *arc*AB?
e) What is the measure of *arc*ABC?

6) Perform the following proof.

Given: Vertical angles ∠AOB and ∠COD are central angles of both circles.

Prove: *arc*AB = *arc*CD.

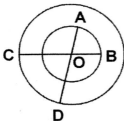

7) Indicate if each of the following are a: *chord, tangent,* or *secant*.

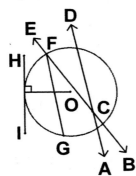

a) HI
b) FG
c) AD
d) BE

8) Copy the following and construct tangents to the circle O from the external point P.

9) Given the circle O and chords GF and IH, find JF to the nearest hundredth.

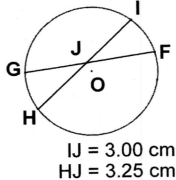

IJ = 3.00 cm
HJ = 3.25 cm
GJ = 2.95 cm
JF = ?

10) Given secant lines BD and CE, answer the following. *Arc*ED = 100° and *arc*BC = 20°.

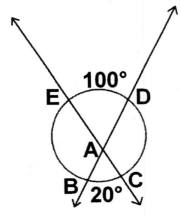

a) Which arc is intercepted by ∠EAD?
b) How many secant angles are formed by BD and CE?
c) What is the measure of ∠BAC?

END

GEOMETRY: Test on Chapter 12   Name_____

1) State whether each of the following are True or False.

   a) A *reflex angle* is an angle whose measure is more than 180°.

   b) Inscribed angles that intercept the same arc are equal.

   c) A secant line is also a radius.

   d) A tangent line intersects a circle in two points.

2) Answer the following using the circle with center B.

   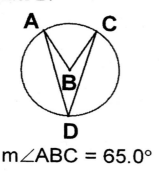

   m∠ABC = 65.0°

   a) Is ∠ABC an inscribed angle?

   b) Is ∠ABC a central angle?

   c) What is the measure of arc AC?

   d) Is AB a chord?

   e) What is the measure of ∠ADC.

3) Perform the following proof.

   Given: ABCD is a trapezoid and arcAB = arcCD.

   Prove: ∠B = ∠C.

   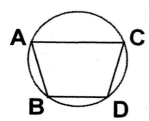

4) Answer the following using circle O and equal chords AB and EF. The measure of arcAB = 100°.

   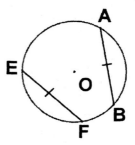

   a) What does the measure of arcEF equal?

   b) If the measure of arcFB is 20°, what is the measure of arcAE?

5) Use the circle to answer the following.

   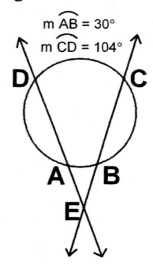

   a) Is ∠DEC a secant angle?

   b) What is the sum of arcAD and arcBC?

   c) What is the measure of ∠DEC?

   TURN OVER

GEOMETRY: Test on Chapter 12   Name_____

6) Use the circle to answer the following. AC is a diameter of the circle.

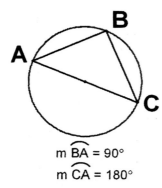

m $\overarc{BA}$ = 90°
m $\overarc{CA}$ = 180°

a) What is the measure of *arc*ABC?

b) What is the measure of *arc*BC?

c) Which chords are equal?

d) Is ∠A an *inscribed* angle?

7) Use the circle and secants to answer the following.

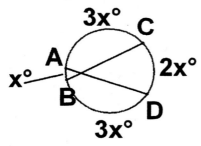

a) What is the value of *x*?

b) What is the measure *arc*AB, *arc*AC, *arc*BD and *arc*CD?

c) What are the measure of the angles formed by the cords?

8) Given AB and BC are radii and AD and DC are tangent to the circle, answer the following.

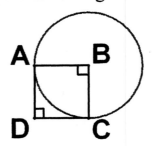

a) Why does AB = BC?

b) Why does DA = DC?

c) Why is BA ⊥ AD and BC ⊥ CD?

9) Describe what the following are.

a) *Semicircle*.

b) *Minor arc*.

c) *Major arc*.

10) Use the circle to determine the measure of *arc*ABC.

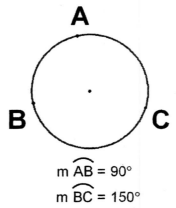

m $\overarc{AB}$ = 90°
m $\overarc{BC}$ = 150°

END

GEOMETRY: Test on Chapter 13   Name_____

1) State whether each of the following are True or False.

   a) The lines containing the altitudes of a triangle are concurrent.

   b) Every triangle is not cyclic.

   c) The medians of a triangle are not concurrent.

   d) A quadrilateral is cyclic iff a pair of its opposite angles are supplementary.

2) Use the △ABC, whose vertices are contained by circle O, to answer the following.

   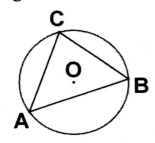

   a) Is △ABC *cyclic*?

   b) Is △ABC *inscribed* in circle O?

   c) Is △ABC *circumscribed* about circle O?

   d) Is point O the *circumcenter* of △ABC?

3) Perform the following construction.

   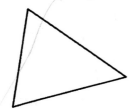

   a) Copy the above triangle.

   b) Circumscribe a circle about the triangle.

4) Which of the following quadrilaterals are cyclic?

   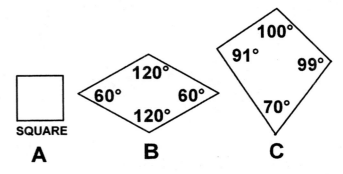

   SQUARE   A   B   C

5) Use the inscribed circle, △ABC and its angle bisectors to answer the following.

   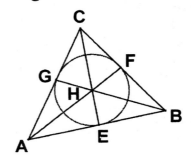

   a) Are the angle bisectors concurrent?

   b) What segments shown are equal to *r*, the radius of H?

   c) Is △ABC inscribed in circle H?

   d) Is circle H the *incircle* of △ABC?

TURN OVER

13A-1

GEOMETRY: Test on Chapter 13   Name_____

6) Use △ABC and its *medians* to answer the following.

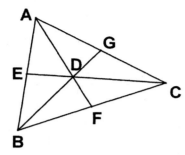

  a) Which line segments are the *medians*?

  b) Are the medians concurrent?

  c) Which point is the *centroid*?

  d) Does AG = GC?

7) Use △ABC and the lines containing its *altitudes* to answer the following.

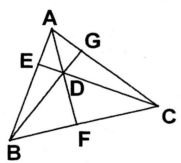

  a) Are the lines containing the altitudes concurrent?

  b) Which point is the *orthocenter* of △ABC?

  c) What is the measure of ∠BGA?

  d) Why is quadrilateral AEDG cyclic?

8) Use the triangle and the concurrent *cevians* to complete the formula.

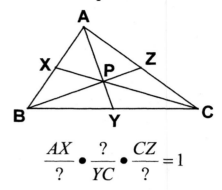

$$\frac{AX}{?} \cdot \frac{?}{YC} \cdot \frac{CZ}{?} = 1$$

9) Use the triangle and the concurrent *cevians* to determine the value of x.

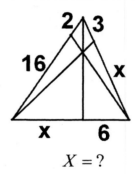

$X = ?$

10) To find the center of a cirlce inscribed in a triangle, first construct two _____ of the triangle?

END

GEOMETRY: Test on Chapter 13   Name_____

1) State whether each of the following are True or False.

   a) The perpendicular bisectors of the sides of a triangle are concurrent.

   b) Every right triangle is cyclic.

   c) The lines containing the altitudes of a triangle are concurrent.

   d) The *centroid* of a triangle is the point in which its medians are concurrent.

2) Which of the following quadrilaterals are cyclic?.

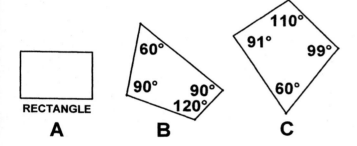

3) Perform the following construction.

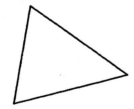

   a) Copy the above triangle.
   b) Inscribe a circle in the triangle.

4) Use △ABC and △HIJ to answer the following.

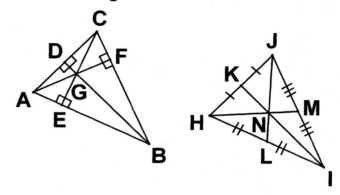

   a) Which of the line segments are altitudes?

   b) Which of the line segments are medians?

   c) Which point is a *centroid*?

   d) Which point is an *orthocenter*?

5) Use the triangle and the concurrent *cevians* to complete the formula

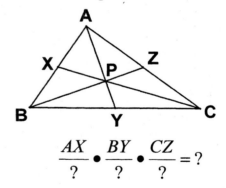

$$\frac{AX}{?} \cdot \frac{BY}{?} \cdot \frac{CZ}{?} = ?$$

TURN OVER

13B-1

GEOMETRY: Test on Chapter 13   Name_____

6) Use right triangle $\triangle ABC$ and the fact that, BC = 8, CA = 6, CY = CZ = 2.

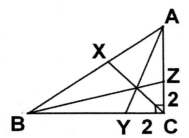

a) What does $\dfrac{BY}{YC}$ equal?

b) What does $\dfrac{CZ}{ZA}$ equal?

c) What does $\dfrac{AX}{XB}$ equal?

d) What does AX equal?

e) What does XB equal?

7) Use equilateral triangles A, B and C to answer the following.

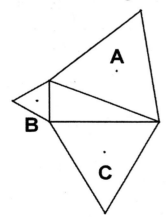

What type of triangle would be formed, if the centers of triangles A, B and C where joined by line segments?

8) State whether each of the following are True or False.

a) Every triangle has an *incircle*.

b) The *medians* of a triangle are concurrent.

c) An equilateral triangle is *not* cyclic.

d) All *cevians* of a triangle must be concurrent.

9) Perform the following construction.

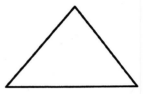

a) Copy the triangle.

b) Circumscribe a circle about the triangle

10) Why are the cevians of the following triangle not concurrent?

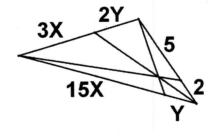

END

GEOMETRY: Test on Chapter 14   Name_____

1) State whether each of the following are True or False.

   a) A *regular polygon* is a convex polygon that is neither equilateral nor equiangular.

   b) Every regular polygon is cyclic.

   c) An *apothem* of a regular polygon is a line segment from its center to the vertex of one of its angles.

   d) If the diameter of a circle is d, its circumference is πd.

2) Use the regular polygon inscribed in the circle to answer the following.

   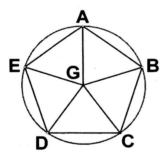

   a) How many *apothems* are shown?

   b) What is the name of the inscribed polygon?

   c) If AB = 2 cm, what is the perimeter of the polygon?

   d) What is the measure of *arc*DC?

3) If the perimeter of an octagon is 40 inches, what is the length of each of its sides?

   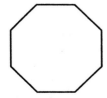

   P = 40 inches

4) Use the regular polygons with radii r to answer the following.

   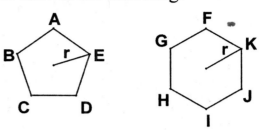

   a) If the radius of the pentagon is 10 meters, what is the perimeter of the pentagon to the nearest hundredth?

   b) If the radius of the hexagon is 10 meters, what is the perimeter of the hexagon?

   c) If the radius of the pentagon is 100 meters, what is the perimeter of the pentagon to the nearest tenth?

   d) If the radius of the hexagon is 100 meters, what is the perimeter of the hexagon?

5) Use the regular polygons with radii r to answer the following.

   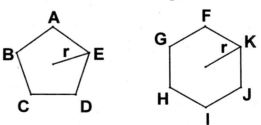

   a) If the radius of the pentagon is 10 meters, what is the area of the pentagon to the nearest hundredth?

   b) If the radius of the hexagon is 100 meters, what is the area of the hexagon?

TURN OVER

GEOMETRY: Test on Chapter 14   Name_____

6) Fill in the blank to make the following statements true.

   a) The perimeter of a regular polygon having *n* sides is 2Nr, in which N = _____ and *r* is its radius.

   b) The area of a regular polygon having *n* sides is Mr^2, in which M = _____ and *r* is its radius.

   c) The *circumference* of a circle is the _____ of the perimeters of the inscribed regular polygons.

   d) The radius of a circle is *r*, its circumference is _____.

7) Use the semicircles centered at B, D, and E, to answer the following.

   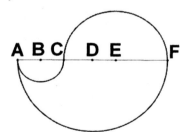

   a) What is the exact length of semicircle B, if the radius AB = 2 meters?

   b) What is the exact length of semicircle D, if the diameter AF = 8 meters?

   c) What is the exact length of semicircle E, if the radius EF = *x* meters?

   d) What is the exact length of semicircle E, if the diameter CF = *y* meters?

8) Fill in the blank to make the following statements true.

   a) The *area* of a circle is the _____ of the areas of the inscribed regular polygons.

   b) If the radius of a circle is *r*, its area is _____.

   c) A *sector* of a circle is a region bounded by an arc of the circle and the two _____ to the endpoints of the arc.

9) Use the figure below to answer the following.

   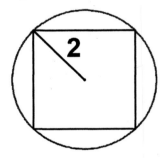

   a) What is the area of the square?

   b) What is the exact area of the circle?

   c) What is the exact area of the region inside the circle that is **not** shaded?

10) What is the exact area of the sector AOB of a circle with radii 30 inches and an arc of 120°?

   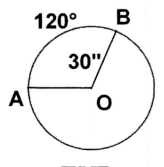

END

GEOMETRY: Test on Chapter 14   Name_____

1) State whether each of the following are True or False.

   a) If the radius of a circle is *r*, its circumference is $2\pi r$.

   b) Any eight sided figure is a *regular octagon*.

   c) The perimeter of a polygon can be found by multiplying the length of a side by the number of sides.

   d) The radius of a regular polygon always equals the length of one of its sides.

2) Use the regular polygon inscribed in the circle to answer the following.

   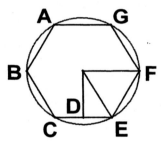

   a) How many *apothems* are shown?

   b) What is the name of the polygon?

   c) If AB = 3 cm, what is the perimeter of the polygon?

   d) What is the measure of *arc*EF?

3) If the perimeter of an octagon is 80 inches, what is the length of each of its sides?

   P = 80 inches

4) Use the regular polygons with radii *r* to answer the following.

   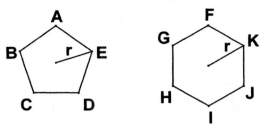

   a) If the radius of the hexagon is 5 meters, what is the perimeter of the hexagon?

   b) If the radius of the pentagon is 5 meters, what is the perimeter of the pentagon to the nearest hundredth?

   c) If the radius of the pentagon is 50 meters, what is the perimeter of the pentagon to the nearest tenth?

   d) If the radius of the hexagon is 50 meters, what is the perimeter of the hexagon?

5) Use the regular polygons with radii *r* to answer the following.

   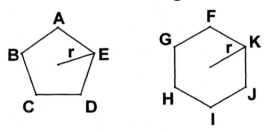

   a) If the radius of the hexagon is 5 meters, what is the area of the hexagon to the nearest hundredth?

   b) If the radius of the pentagon is 50 meters, what is the area of the pentagon?

TURN OVER   14B-1

GEOMETRY: Test on Chapter 14   Name_____

6) Fill in the blank to make the following statements true.

   a) The perimeter of a regular polygon having n sides is $2Nr$, in which _____ $= n\sin\dfrac{180}{n}$ and r is its radius.

   b) The area of a regular polygon having n sides is Mr^2, in which r is its radius and _____ $= n\sin\dfrac{180}{n}\cos\dfrac{180}{n}$.

   c) The _____ of a circle is the limit of the perimeters of the inscribed regular polygons.

   d) The radius of a circle is r, its _____ is $2\pi r$.

7) Use the circles centered at B, D, and E, with radii AB, DF, and EF respectively, to answer the following.

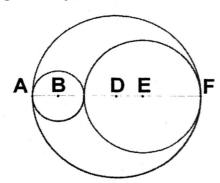

   a) What is the exact area of circle D, if AB = 2 cm and EF = 4 cm?

   b) What is exact area of circle B, if AF = 10 cm and EF = 4 cm?

   c) What is the exact area of the non-shaded region, if AB = 2 cm and AF = 12 cm?

8) Fill in the blank to make the following statements true.

   a) The _____ of a circle is the limit of the areas of the inscribed regular polygons.

   b) If the radius of a circle is r, its _____ is πr^2.

   c) A _____ of a circle is a region bounded by an arc of the circle and the two radii to the endpoints of the arc.

9) Use the figure below to answer the following.

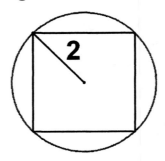

   a) What is the area of the square?

   b) What is exact are of the circle?

   c) What is the exact length of $arc AB$?

10) What is the exact area of the sector AOB with radii 10 inches and an arc of 120°?

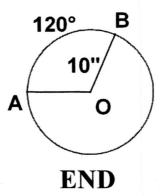

END

GEOMETRY: Test on Chapter 15   Name_____

1) State whether each of the following are True or False.

   a) Two lines are *skew* iff they are not parallel and do intersect.

   b) A *regular polyhedron* is a solid whose faces are congruent regular polygons and in which an equal number of polygons meet at each vertex.

   c) A *cross section* of a geometric solid is the intersection of a plane and the solid.

   d) An *altitude* of a prism is the length of a line segment that connects the planes of its bases and is perpendicular to both of them.

2) Use the regular solid to answer the following.

   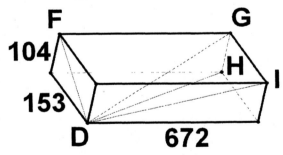

   a) What is the length of line segment DF?

   b) What is the length of line segment DG?

   c) What is the length of line segment DH, to the nearest tenth?

   d) What is the length of line segment DI?

3) Use the prism to answer the following.

   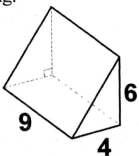

   a) What shape are the lateral faces?

   b) What shape are the bases?

   c) What is the altitude?

   d) What is the volume?

4) The following is a regular pyramid whose base is a pentagon with a radius of 5 cm. If the altitude is 15 cm, what is the volume of the pyramid to the nearest hundredth?

   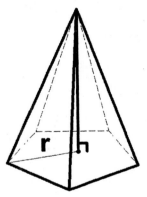

5) Answer the following.

   a) What is the exact volume of an oblique cylinder with a radius of 5 inches and an altitude of 12 inches?

   b) What is the exact volume of a right cone with a radius of 10 inches and a altitude of 30 inches?

TURN OVER   15A-1

GEOMETRY: Test on Chapter 15   Name_____

6) Fill in the blank to make the following statements true.

   a) The number of faces an octahedron has is _____.

   b) The number of faces a tetrahedron has is _____.

   c) If the ratio of a pair of corresponding dimensions of two similar solids is r, then the ratio of their volumes is _____.

   d) A _____ is the set of all points in space that are at a given distance from a given point.

7) Use a sphere of radius 4 meters to answer the following.

   a) What is the exact surface area of the sphere?

   b) What is the surface area of the sphere to the nearest hundredth?

   c) What is the exact volume of the sphere?

   d) What is the volume of the sphere to the nearest hundredth?

8) Indicate whether each of the following solids are always similar. If not similar, sketch an example of why.

   a) Two cones.

   b) Two cubes.

   c) Two square pyramids.

   d) Two spheres.

9) A cube and a tetrahedron are inscribed in the same sphere. Which has an area and volume closer to that of the sphere, the cube or the tetrahedron?

10) The ratio of the corresponding dimensions of two cubes is:
$$\frac{3x}{x} = 3$$

   a) What is the ratio of their surface areas?

   b) What is the ratio of their volumes?

END

GEOMETRY: Test on Chapter 15   Name_____

1) State whether each of the following are True or False.

   a) If two points lie in a plane, the line that contains them lies in the plane.

   b) If two planes intersect, they intersect in a line.

   c) If the ratio of a pair of corresponding dimensions of two similar solids is r, then the ratio of their surface areas is r^3.

   d) The surface area of a sphere is four times the product of π and its radius. $A = 4\pi r$.

2) Use the regular solid to answer the following.

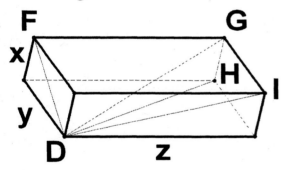

   a) What is the length of line segment DF in terms of x and y?

   b) What is the length of line segment DG in terms of x, y and z?

   c) What is the length of line segment DH in terms of y and z?

   d) What is the length of line segment DI in terms of x and z?

3) Use the prism to answer the following.

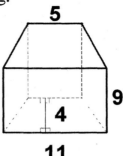

   a) What shape are the lateral faces?

   b) What shape are the bases?

   c) What is the altitude?

   d) What is the volume?

4) Given a regular pyramid whose base is a pentagon with a radius of 10 cm and an altitude of 25 cm, what is the volume of the pyramid to the nearest hundredth?

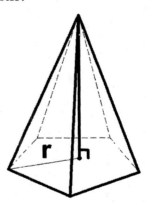

5) Answer the following.

   a) What is the exact volume of an oblique cylinder with a radius of 5 inches and an altitude of 10 inches?

   b) What is the exact volume of a right cone with a radius of 2 inches and a altitude of 12 inches?

TURN OVER   15B-1

GEOMETRY: Test on Chapter 15   Name_____

6) Fill in the blank to make the following statements true.

   a) The number of faces an _____ has is eight.

   b) The number of faces a _____ has is four.

   c) If the ratio of a pair of corresponding dimensions of two similar solids is r, then the ratio of their _____ is r^3.

   d) A sphere is the set of all points in space that are at a given distance from a given _____.

7) Use a sphere of radius 3 centimeters to answer the following.

   a) What is the exact surface area of the sphere?

   b) What is the surface area of the sphere to the nearest hundredth?

   c) What is the exact volume of the sphere?

   d) What is the volume of the sphere to the nearest hundredth?

8) Indicate whether each of the following solids are always similar. If not similar, sketch an example of why.

   a) Two cones.

   b) Two cubes.

   c) Two square pyramids.

   d) Two spheres.

9) An octahedron and an icosahedron are inscribed in the same sphere. Which has an area and volume closer to that of the sphere, the octahedron or the icosahedron?

10) The ratio of the corresponding dimensions of two cubes is:
$$\frac{5x}{x} = 5$$

   a) What is the ratio of their surface areas?

   b) What is the ratio of their volumes?

END

GEOMETRY: Test on Chapter 16   Name_____

1. State whether each of the following are True or False.

   a) A *great circle* of a sphere is a set of points that is the intersection of the sphere and a plane containing its center.

   b) The summit angles of a Saccheri quadrilateral are not equal.

   c) If the legs of a birectangular quadrilateral are unequal, the summit angles opposite them are unequal in the same order.

   d) The line segment joining the midpoints of the base and summit of a Saccheri quadrilateral is perpendicular to both of them.

2. Use the following figure to determine whether each of the following seems to be true in **both** *Euclidean* and *sphere* geometry. If only true in one, name that geometry.

   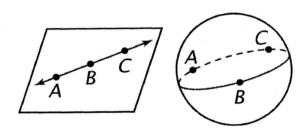

   a) Points A, B and C are collinear.

   b) If two points lie in a plane, the line that contains them lies in the plane.

   c) Point B is between points A and C.

   d) Point B separates the line into two parts.

3. The following sphere is divided by great circles into congruent triangles. Each triangle is a 36°-60° right triangle.

   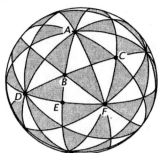

   a) What is the sum of the angles of one of the congruent triangles?

   b) Explain why the acute angles of these right triangles are not complementary.

   c) What is the measure of ∠AEF?

   d) What is the measure of ∠EFA?

   e) What is the measure of ∠FAE?

   f) Is △ADC a right triangle?

   g) What is the sum of the measures of △ABC?

4. Answer the following using spherical geometry.

   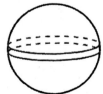

   a) How many great circles appear on the sphere?

   b) Why is there actually only one line on the sphere?

   c) Why do parallel lines not exist on a sphere?

TURN OVER   16A-1

GEOMETRY: Test on Chapter 16   Name_____

5. Complete the following proof by providing the reasons for each step.

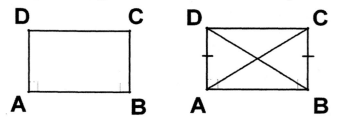

Given: Saccheri quadrilateral ABCD with base AB.
Prove: ∠C = ∠D.

a) Draw AC and BD. Why?
b) AD = BC. Why?
c) ∠BAD = ∠ABC. Why?
d) AB = AB. Why?
e) △BAD ≅ △ABC. Why?
f) BD = AC. Why?
g) ∠CAB = ∠DBA. Why?
h) ∠DAC = ∠CBD. Why?
i) △BCD ≅ △ADC. Why?
j) ∠BCD = ∠ADC. Why?

6. Complete the statement for each Geometry listed.

a) In Euclidean Geometry:
The summit angles of a Saccheri quadrilateral are _____ angles.

b) In Lobachevsky Geometry:
The summit angles of a Saccheri quadrilateral are _____ angles.

c) In Riemann Geometry:
The summit angles of a Saccheri quadrilateral are _____ angles.

7. Use quadrilateral ACDF to answer the following in *Lobachevskian* geometry.

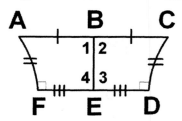

AF ⊥ FD, CD ⊥ FD, AB = BC,
FE = ED and AF = CD

a) What kind of quadrilateral is ACDF?.

b) What kind of angles are ∠A and ∠C?

c) Why are ∠1, ∠2, ∠3 and ∠4, right angles?

8. Answer the following true or false.

a) In Lobachevskian geometry, the sum of the angles of a triangle is less than 180°.

b) In Lobachevskian geometry, the sum of the angles of a quadrilateral is 360°.

9. Explain why, in Lobachevskian geometry, the acute angles of a right triangle are not complementary.

10. What is the definition of a *birectangular quadrilateral*?

END

GEOMETRY: Test on Chapter 16   Name_____

1. State whether each of the following are True or False.

   a) A *Saccheri quadrilateral* is a birectangular quadrilateral whose legs are not equal.

   b) A *birectangular quadrilateral* is a quadrilateral that has two sides perpendicular to a third side.

   c) *Lobachevskian geometry* is a non-Euclidean geometry.

   d) In *Riemannian geometry*, the sum of the angles of a triangle is the same as in Euclidean geometry.

2. Use the Equilateral Right triangle and the sphere of radius 4 units to answer the following.

   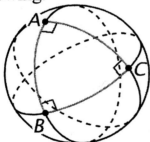

   a) A great circle of the sphere, is how many times greater than leg BC?

   b) The length of leg BC is what fraction of a greater circle of the sphere?

   c) What is the length of the line BC?

   d) What is the perimeter of △ABC?

   e) What is the surface area of the sphere?

   f) What is the area of △ABC?

3. The following sphere is divided by great circles into congruent triangles. Each triangle is a 36°-60° right triangle.

   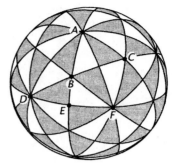

   a) What is the sum of the angles of one of the congruent triangles?

   b) What type of geometry is being used here?

   c) What is the measure of ∠ABC?

   d) What is the measure of ∠EFC?

   e) What is the measure of ∠BEF?

   f) Is △AEF a right triangle?

   g) Is △AEF a 36°- 60° right triangle?

4. Answer the following using spherical geometry.

   a) How many great circles does any sphere contain?

   b) Does every great circle of a sphere intersect all other great circles of the sphere?

   c) Can two great circles be parallel to one another?

   d) What does a great circle divide the sphere into?

TURN OVER

16B-1

GEOMETRY: Test on Chapter 16   Name_____

5. Complete the following proof by providing the reasons for each step.

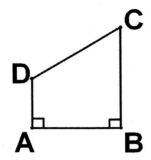 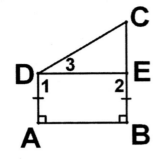

Given: Birectangle quadrilateral ABCD with base AB,

DA ⊥ AB and CB ⊥ AB
and CB > DA.

Prove: ∠D = ∠C.

a) Choose a point E on BC so that BE = AD. Why?

b) Draw DE. Why?

c) ABED is a Saccheri quadrilateral. Why?

d) ∠1 = ∠2. Why?

e) Because ∠ADC = ∠1 + ∠3, ∠ADC > ∠1. Why?

f) ∠ADC > ∠2 Why?

g) ∠2 > ∠C. Why?

h) ∠ADC > ∠C. Why?

6. Complete the statement for each geometry listed using either *no, exactly one,* or *more than one.*

a) In Euclidean Geometry: Through a point not on a line there is _____ parallel to the line.

b) In Lobachevsky Geometry: Through a point not on a line there is _____ parallel to the line.

c) In Riemann Geometry: Through a point not on a line there is _____ parallel to the line.

7. Use the Riemannian Quadrilateral to answer the following.

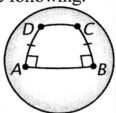

a) Are ∠D and ∠C acute, right or obtuse?.

b) Whose length is greatest, DC or AB?

c) In Riemann Geometry, is the summit of a Saccheri quadrilateral shorter than its base?

8. Answer the following true or false.

a) In Lobachevskian geometry, the sum of the angles of a triangle is 180°.

b) In Lobachevskian geometry, If two triangles are similar, they must also be congruent.

9. Explain why, in Lobachevskian geometry, that if two angles of two triangle are equal, then the third angles are not necessarily equal.

10. What is the definition of *polar points*?

END

GEOMETRY: Mid-Term Test   Name_____

1) State whether each of the following are true or false.

   a) A *point* can be described as "that which has no part".

   b) A *line segment* is not bounded.

   c) A *line* and a *line segment* are the same thing.

   d) An *angle* is a pair of rays that have the same endpoint.

2) Write the "if-then" statement represented by the following Euler diagram.

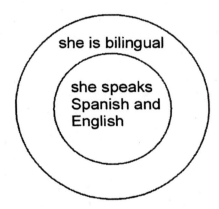

3) Indicate which property of equality is illustrated by each of the following statements.

   a) If $x + y = x + z$, then $y = z$.

   b) If $d = 2r$, then $\dfrac{d}{2} = r$.

   c) If $a = b$ and $a + c = 10$, then $b + c = 10$.

   d) If $\angle A + \angle B + \angle C - 90° = 90°$, then $\angle A + \angle B + \angle C = 180°$.

4) **Given**: equilateral $\triangle ABC$ and the fact that BD bisects $\angle ABC$.
   Prove: $\triangle ABD \cong \triangle CBD$ (using SAS)

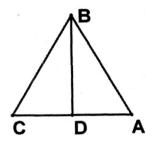

5) Answer the following based on the figure.

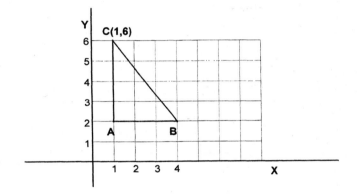

   a) What are point A's coordinates?

   b) What are point B's coordinates?

   c) What is the distance between points A and B?

   d) What is the distance between points A and C?

   e) What is the distance between points C and B?

6) Given $\triangle ABC$ (not shown here), such that AB = 3.0 cm, BC = 3.15 cm and CA = 2.0 cm. Which angle is the least: $\angle A$, $\angle B$ or $\angle C$?

TURN OVER

Mid-Term A-1

GEOMETRY: Mid-Term Test

Name_____

7) Fill in the blank using information from the following figure.

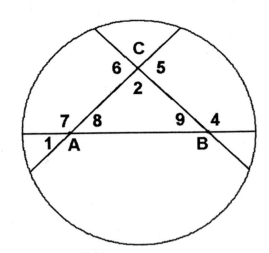

a) AB + BC > _____.

b) ∠4 + ∠9 = _____ degrees.

c) ∠2 + ∠8 + ∠9 = _____ degrees.

d) ∠4 = ∠____ + ∠____.

e) ∠7 > ∠____ and ∠7 > ∠____.

8) Use a straight edge and compass to perform the following construction.

_____ *j*

a) Copy line segment *j*.

b) Find the midpoint M of *j*.

c) Construct a line perpendicular to *j*, through point M

9) Fill in the blank to make the definition correct.

a) A _____ is a quadrilateral all of whose sides and angles are equal.

b) A _____ is a quadrilateral all of whose sides are equal.

c) A _____ is a quadrilateral all of whose angles are equal.

d) If the ratio of a pair of corresponding dimensions of two similar solids is *r*, then the ratio of their _____ is r^3.

10) Identify the correct transformation: *translation* or *glide reflection*.

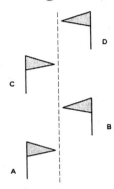

a) Flag A to image C.

b) Flag A to image B.

c) Flag D to image A.

d) Flag D to image B.

e) Flag D to image C.

END

Mid-Term A-2

GEOMETRY: Mid-Term Test   Name_____

1. Use a protractor and/or the triangle below answer the following:

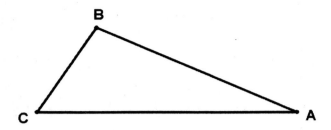

   a) What is the measure of ∠A?
   b) What is the measure of ∠B?
   c) What is the measure of ∠C?
   d) Are ∠B and ∠C complementary angles?
   e) Is ∠A, a right angle?

2. Consider the statement,
   "If you are 18, then you can vote."

   a) What is the hypothesis?
   b) What is the conclusion?
   c) In symbolic form, a → b, what is *a* and what is *b*?
   d) What is the *converse*?
   e) Is the converse a conditional statement?

3. Answer the following based on the figure below.

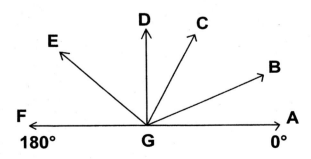

   a) Is ∠AGE acute?
   b) Are ∠AGD and ∠FGD Supplementary?
   c) Is ray GB between ray GA and ray GE?
   d) What is the measure of ∠BGE, if ∠AGB = 40° and ∠AGE = 135°?
   e) Write the angular equation that follows from the fact that GC-GD-GE.
   f) If ray GE bisects ∠DGF, does ∠DGE = ∠EGF?
   g) Is ∠AGF a straight angle?

4. **Given:** square ABCD and the fact that AC bisects ∠DAB and ∠BCD.
   Prove: △ABC ≅ △ADC (using ASA)

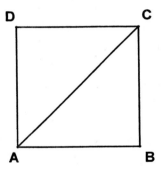

5. Fill in the blanks to complete the following definition of congruent.

   Two triangles are *congruent* iff there is a correspondence between their vertices such that all of their corresponding _____ and _____ are equal.

TURN OVER

Mid-Term B-1

GEOMETRY: Mid-Term Test   Name_____

6. Use the figure below to answer the following.

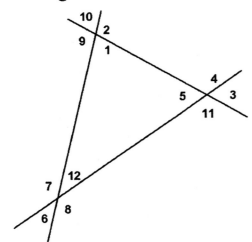

   a) Which angles are exterior angles of the triangle?

   b) Which is greater ∠1 or ∠12?

   c) If ∠10 = 100°, can ∠11 = 80°?

7. Answer the following based on the parallel lines and transversal below.

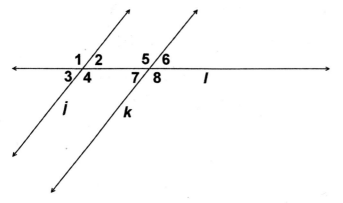

   a) List all pairs of *corresponding angles*.

   b) List all pairs of *alternate interior angles*.

   c) List all pairs of *interior angles on the same side of the transversal*.

8. Perform the following construction.

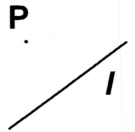

   a) Copy the line *l* and point P above.

   b) Construct a line parallel to *l*, through point P.

9. Given a quadrilateral ABCD, identify each as square, rectangle or rhombus.

   a) AB = BC = CD = DA

   b) AB = BC = CD = DA and ∠A = ∠B = ∠C = ∠D

   c) ∠A = ∠B = ∠C = ∠D

10. Fill in the blank to complete the definition.

    a) A _____ is the composite of two successive reflections through intersecting lines.

    b) A _____ is the composite of a translation and a reflection in a line parallel to the direction of the translation.

    c) Two figures are _____ if there is an isometry such that one figure is the image of the other.

END

GEOMETRY: Final Test   Name_____

1. State whether each of the following are True or False.

   a) The area of a right triangle is one half the product of its legs.

   b) If two angles of one triangle are equal to two angles of another triangle, the triangles are similar.

   c) Two non-vertical lines are parallel iff their slopes are equal.

   d) If the diameter of a circle is d, its circumference is πd.

2. What is the area of the following triangle?

   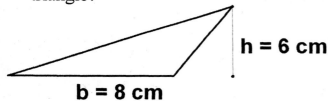

3. Use the two triangles, labeled as shown, to answer the following using "Yes" or "No".

   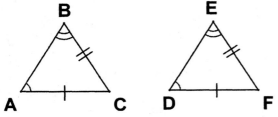

   a) Are the triangles congruent by ASA?

   b) Are the triangles congruent by SAS?

   c) Are the triangles congruent by SSS?

   d) Are the triangles congruent by AA?

4. Given parallel lines l and j, cut by a transversal, answer the following.

   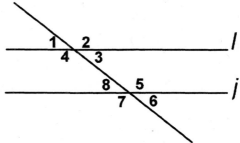

   a) List all of the angles that are equal to $\angle 1$.

   b) List all of the angles supplementary to $\angle 1$.

5. Use the figure to answer the following. BE // CD.

   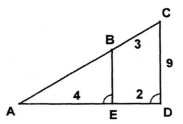

   a) Why is $\angle AEB = \angle ADC$?

   b) Why is $\triangle ABE \sim \triangle ACD$?

   c) What is the length of BE?

   d) What is the ratio of the perimeter of $\triangle ABE$ to the perimeter of $\triangle ACD$?

6. Perform the following construction.

   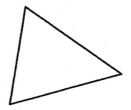

   a) Copy the above triangle.

   b) Circumscribe a circle about the triangle.

TURN OVER   Final A-1

GEOMETRY: Final Test   Name_____

7. Use the regular polygons with radii *r* to answer the following.

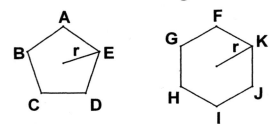

   a) If the radius of the pentagon is 10 meters, what is the perimeter of the pentagon to the nearest hundredth?

   b) If the radius of the hexagon is 10 meters, what is the perimeter of the hexagon?

   c) If the radius of the pentagon is 100 meters, what is the perimeter of the pentagon to the nearest tenth?

   d) If the radius of the hexagon is 100 meters, what is the perimeter

8. Using an octagon with equal sides, equal angles and a perimeter of 80 inches, answer the following.

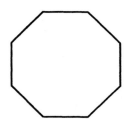

P = 80 inches

   a) Is the octagon cyclic?

   b) Is the octagon regular?

   c) How many sides are there?

   d) What is the length of each of its sides?

9. Use a sphere of radius 9 meters to answer the following.

   a) What is the exact surface area of the sphere?

   b) What is the surface area of the sphere to the nearest hundredth?

   c) What is the exact volume of the sphere?

   d) What is the volume of the sphere to the nearest hundredth?

10. State whether each of the following are True or False.

   a) In Euclidean geometry, through a point not on a line there is exactly one parallel to the line.

   b) On a sphere, the summit angles of a Saccheri quadrilateral are acute.

   c) On a sphere, the summit of a Saccheri quadrilateral is longer than its base.

   d) In Riemann geometry, through a point not on a line there is no parallel to the line.

END

GEOMETRY: Final Test   Name_____

1. State whether each of the following are True or False.

   a) Two triangles similar to a third triangle are not similar to each other.

   b) The slope of a line is found by dividing the rise by the run.

   c) A tangent line intersects a circle in only one point.

   d) No regular polygon is cyclic.

2. **Given**: In △ABC, AB = AC and DE // BC.

   Prove: AD = AE

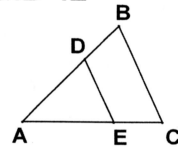

3. Use the figure to answer the following. ∠AEB = ∠ADC.

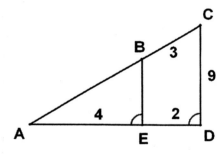

   a) Why is BE // CD?

   b) Why is △ABE ~ △ACD?

   c) What is the length of BE?

   d) What is the length of AB?

   e) What is the ratio of the area of △ACD to the area of △ABE?

4.
   a) Draw a figure which has point symmetry.

   b) Does your figure also have line symmetry?

   c) If you were to reflect your figure through a line and then through another line perpendicular to the first line, what single isometry would this composite of the two reflections be equal to?

5. Perform the following construction.

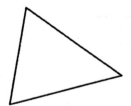

   a) Copy the above triangle.

   b) Inscribe a circle in the triangle.

6. Use the circle to answer the following. AC is a diameter of the circle.

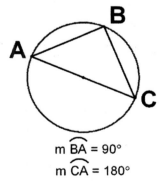

   m $\overarc{BA}$ = 90°
   m $\overarc{CA}$ = 180°

   a) What is the measure of *arc*ABC?

   b) What is the measure of *arc*BC?

   c) Which chords are equal?

   d) What is the measure of ∠A?

   TURN OVER

   Final B-1

GEOMETRY: Final Test

Name_____

7. Use the regular polygons with radii r to answer the following.

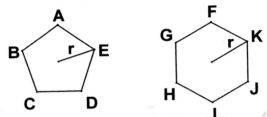

   a) If the radius of the hexagon is 5 meters, what is the perimeter of the hexagon?

   b) If the radius of the pentagon is 5 meters, what is the perimeter of the pentagon to the nearest hundredth?

   c) If the radius of the pentagon is 50 meters, what is the perimeter of the pentagon to the nearest tenth?

   d) If the radius of the hexagon is 50 meters, what is the perimeter of the hexagon?

8. Use a sphere of radius 3 centimeters to answer the following.

   a) What is the exact surface area of the sphere?

   b) What is the surface area of the sphere to the nearest hundredth?

   c) What is the exact volume of the sphere?

   d) What is the volume of the sphere to the nearest hundredth?

9. Use the prism to answer the following.

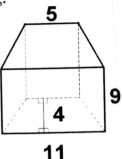

   a) What shape are the lateral faces?

   b) What shape are the bases?

   c) What is the altitude?

   d) What is the volume?

10. Explain the following.

    a) Why a *Saccheri quadrilateral* is also a *birectangle quadrilateral*.

    b) Why in Spherical Geometry, through a point not on a line, there are *no lines parallel to the line*.

END

GEOMETRY: *Answers to* Chapter 1

1. a) True
   b) False
   c) False
   d) True

2. a) ...all of them
   b) ...the same point
   c) ...contains all of them

3. Sketch should be a line containing three points.

4. a) approximately 24° (protractors vary)
   b) approximately 101°
   c) approximately 56°

5. a) quadrilateral
   b) pentagon
   c) octagon
   d) hexagon
   e) triangle

6. a) 9 square inches
   b) 12 inches
   c) Yes

7. A bisected line should be constructed.

8. A bisected angle should be constructed.

9. Each bisected angle would be 45 degrees.

10. Three dimensional

1A-1

GEOMETRY: *Answers to* Chapter 1

1. a) Any 4 of CG, CD, GD, FG, GE, FE, HG, CK, GK, KD
   b) *l* or HG
   c) C, G, K, D
   d) Any 3 w/out line through all (ex. C, G and H)

2. a) BC and BD
   b) Draw a line segment CD, Approximately 3.6 cm
   c) Approximately 45° (Protractors vary)

3. a) Three
   b) One
   c) Two

4. a) 4 in^2
   b) 8 inches
   c) Yes

5. a) quadrilateral
   b) pentagon
   c) octagon

6.

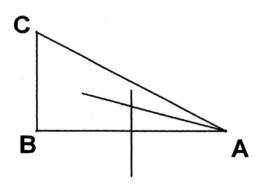

7. Use a compass to draw a circle with a diameter of 4 inches.

8. a) Six
   b) Three

9. Each bisected angle would be 22.5 degrees.

10. AC, BC, DC

GEOMETRY: *Answers to* Chapter 2

1. a) If rain is falling
   b) The clouds are gray
   c) Rain is falling, is a.
      The clouds are gray, is b.

2. a) If the new baby girl is smiling, then she is happy.
   b) If you clean your room, then you can go to the concert.
   c) If your score is the highest, then you win the game.
   d) If the Giant Panda is hungry, then it searches for bamboo.

3. a) If a line is a horizontal line, then it is perpendicular to a vertical line.
   b) If a line is perpendicular to a vertical line, then it is a horizontal line.
   c) Yes, since this is a definition.

4. a) True
   b) True
   c) False
   d) False

5. a) If you live in New York City, then you live in New York State.
      If you live in New York State, then you live in the United States.
      If you live in the United States, then you live in North America.
   b) If you live in New York City, then you live in North America.
   c) Yes, the argument is a syllogism.

6.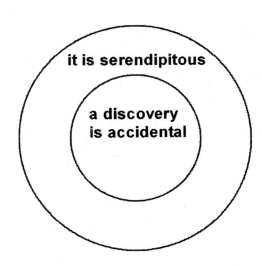

7. If she enjoyed running, she would join the track team.

8. a) Definition
   b) Definition
   c) Postulate
   d) Definition
   e) Postulate

9. Suppose that the sum of two even numbers are ODD (or not even).

10. a) 90°
    b) 60°

GEOMETRY: *Answers to* Chapter 2

1. a) If you eat all of your vegetables
   b) You can have dessert
   c) You eat all of your vegetables, is a.
      You can have dessert, is b.

2. a) If points are collinear, then there is a line that contains all of them.
   b) If you study hard, then you will get good grades.
   c) If your score is the highest, then you win the game.

3. a) If circles are coplanar, then they lie in the same plane or if circles lie in the same plane, then they are coplanar.
   b) Whichever of the above statements was not used above in (a).
   c) Yes, since this is a definition.

4. a) True
   b) False
   c) True
   d) False

5. If this is a direct proof, then this is not an indirect proof.
   If this is not an indirect proof, then there is no contradiction.
   b) If this is a direct proof, then there is no contradiction.
   c) Yes, the argument is a syllogism.

6.

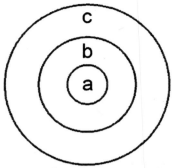

7. A statement that is assumed to be true without proof.

8. a) Postulate
   b) Definition
   c) Definition

9. If she speaks Spanish and English, then she is bilingual.

10. a) 10 cm
    b) 10π cm

GEOMETRY: *Answers to* Chapter 3

1. a) Addition Property
   b) Multiplication Property
   c) Reflexive Property

2. a) Subtraction Property
   b) Division Property
   c) Substitution Property
   d) Addition Property

3. AB + BC = AC

4. $a > b > c$

5. a) True
   b) True
   c) True
   d) False
   e) True

6. a) No
   b) Yes
   c) Yes
   d) 95°
   e) ∠CGD + ∠DGE = ∠CGE
   f) Yes, if a ray bisects an angle, the angle is divided into two equal angles.
   g) Yes

7. Yes

8. a) Yes
   b) Complements of the same angle are equal.

9. a) Vertical angles are equal.
   b) 75°
   c) 60°
   d) No
   e) ∠3 and ∠12.
   f) Yes

10. a) No
    b) Yes
    c) No

3A-1

GEOMETRY: *Answers to* Chapter 3

1. a) Subtraction Property
   b) Division Property
   c) Reflexive Property

2. a) Multiplication Property
   b) Subtraction Property
   c) Addition Property
   d) Substitution Property

3. $a > b > c$

4. $= AC$

5. a) True
   b) False
   c) True
   d) False
   e) True

6. a) Yes
   b) No
   c) Yes
   d) 60°
   e) ∠AFE
   f) Yes
   g) Yes, ∠AFC and ∠EFC

7. 0°

8. a) Because they are not supplementary.
   b) Because they are not equal.

9. a) Vertical angles are equal.
   b) 105°
   c) 75°
   d) Because they are a linear pair.
   e) Any two of ∠2, ∠5 or ∠9
   f) Because they intersect.

10. a) No
    b) When ∠A = 90° and ∠B = 90°.
    c) No, parallel lines lie in the same plane.

GEOMETRY: *Answers to* Chapter 4

1. $\sqrt{(x_2 - x_1)^2 + (y_2 - y_1)^2}$

2. a) 3
   b) 4
   c) 5

3. a) (-4, 2)
   b) (0, 5)
   c) (2, 1)
   d) (4, 2)

4. a) Yes
   b) No
   c) No
   d) Yes

5. a) Yes
   b) Yes
   c) Yes
   d) Yes

6. △ABC is equilateral
   (Given)
   AB = BC
   (Def. of equilateral triangle)
   BD bisects ∠ABC
   (Given)
   ∠CBD = ∠ABC
   (Def. of Angle Bisected)
   BD = BD
   (Reflexive)
   △ABB ≅ △CBD
   (SAS)

7. a) False
   b) True
   c) True

8. a) ASA
   b) SAS
   c) If two triangles are congruent to a third triangle, they are congruent to each other.

9. a) If two angles of a triangle are equal, the sides opposite them are equal.
   b) HF (or AB or AC)

10.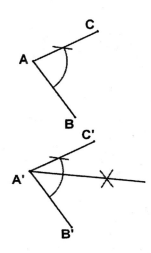

GEOMETRY: *Answers to* Chapter 4

1. a) False
   b) True
   c) True

2. Sides, Angles

3. a) SAS
   b) ASA
   c) If two triangles are congruent to a third triangle, they are congruent to each other.

4. a) Equal
   b) Congruent

5. AC bisects ∠DAB
   (Given)
   ∠DAC = ∠BAC
   (Def. of Angle Bisected)
   AC = AC
   (Reflexive)
   ∠DCA = ∠BCA
   (Def. of Bisected)
   △ABC ≅ △ADC
   (ASA)

6. a) True
   b) False
   c) True
   d) True

7. a) If two sides of a triangle are equal, the angles opposite them are equal.
   b) SSS

8. a) Yes
   b) Yes
   c) 60°

9. A constructed copy of the following triangle.

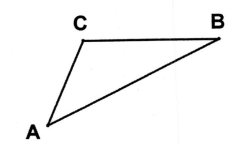

10. a) (1,2).
    b) (4,2)
    c) 3 units
    d) 4 units
    e) 5 units

GEOMETRY: *Answers to* Chapter 5

1. a) Three Possibilities Property
   b) Multiplication Property
   c) Addition Property

2. ...$a + c > b + d$

3. ...$c > a$ and $c > b$

4. a) ∠2, ∠3, ∠5, ∠6, ∠8, ∠9
   b) 65°
   c) No

5. a) False
   b) True
   c) True
   d) True
   e) False

6. ...the angles opposite them are unequal in the same order.

7. ∠Y

8. a) No
   b) Yes

9. a) AC
   b) 180°
   c) 180°
   d) ∠2, ∠8
   e) ∠2, ∠9

10. a) Given
    b) Whole is greater than the part
    c) Transitive Property
    d) If two angles of a triangle are unequal, the sides opposite them are unequal in the same order.

GEOMETRY: *Answers to* Chapter 5

1. a) Transitive Property
   b) Addition Property
   c) Three Possibilities Property

2. $c > d$

3. $a + b = c$

4. a) ∠1, ∠3, ∠5, ∠6, ∠10, ∠12
   b) 100°
   c) No

5. a) False
   b) True
   c) True
   d) True
   e) True

6. …the sides opposite them are unequal in the same order.

7. ∠B

8. a) Yes
   b) No

9. a) AB
   b) 180°
   c) 180°
   d) ∠8, ∠9
   e) ∠2, ∠8

10. a) Given
    b) The triangle angle sum theorem. The sum of the angles of a triangle are 180°.
    c) Substitution Property
    d) Subtraction Property
    e) Substitution Property
    f) The whole is greater than the parts.

g) Theorem 14, if two angles of a triangle are unequal, the sides opposite them are unequal in the same order.

GEOMETRY: *Answers to* Chapter 6

1. a) True
   b) False
   c) False
   d) True

2. a) Line *l*
   b) ∠1 and ∠5, ∠2 and ∠6, ∠3 and ∠7, ∠4 and ∠8
   c) ∠2 and ∠7, ∠4 and ∠5
   d) ∠2 and ∠5, ∠4 and ∠7
   e) Lines *j* and *k*

3. a) Given
   b) Vertical angles are equal.
   c) Substitution
   d) Equal corresponding angles mean that lines are parallel.

4. a) Yes
   b) Yes
   c) No
   d) Yes

5. a) Parallel lines form equal alternate interior angles.
   b) Parallel lines form equal corresponding angles.
   c) Betweeness of rays theorem

6. a) The HL Theorem.
   b) The alternate interior angles are equal (both 90°) or In a plane, two lines perpendicular to a third line are parallel.
   c) An exterior angle of a triangle is equal to the sum of the remote interior angles.

7. a) No
   b) No
   c) Yes
   d) No

8. The sum of the angles of a triangle is 180°.

9.

10. Through a point not on a line, there is exactly one line parallel to the line.

GEOMETRY: *Answers to* Chapter 6

1. a) True
   b) True
   c) True
   d) False

2. a) 45°
   b) 45°
   c) 135°
   d) 135°
   e) Lines *l*

3. a) Given
   b) The angles in a linear pair are supplementary.
   c) Supplements of the same angle are equal.
   d) Equal corresponding angles mean that lines are parallel.

4. a) Yes
   b) No
   c) No
   d) No

5. a) angles are equal.
   b) complementary
   c) 60°

6. a) If two angles of one triangle are equal to two angles of another triangle the third angles are equal.
   b) Each angle of an equilateral triangle is 60°.
   c) The acute angles of a right triangle are complementary.

7. a) The exterior angle of a triangle is equal to the sum of the remote interior angles.
   b) The sum of the angles of a triangle are 180°.

8. If the hypotenuse and a leg of one right triangle are equal to the corresponding parts of another right triangle, the triangles are congruent.

9. a) The Hypotenuse Leg Theorem.
   b) The AAS Theorem.

10.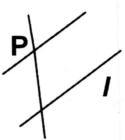

(Several possible methods.)

GEOMETRY: *Answers to* Chapter 7

1. a) True
   b) True
   c) False
   d) False
   e) True

2. a) 10.
   b) Opposite sides of parallelograms are parallel.
   c) 5
   d) 180°

3. a) equal.
   b) equal.
   c) equal.

4. a) True
   b) False
   c) True

5. a) Rhombus
   b) Square
   c) Rectangle

6. a) A quadrilateral that has exactly one pair of parallel sides.
   b) The parallel lines of a trapezoid
   c) A trapezoid whose legs are equal.

7. a) 3 cm
   b) Yes

8. a) AC and BD
   b) 123°
   c) 60°
   d) Yes

9. a) 2 cm
   b) 2x

10. midpoints of its two sides.

GEOMETRY: *Answers to* Chapter 7

1. a) True
   b) False
   c) False
   d) True

2. a) 5 cm.
   b) 40°
   c) 80°
   d) ∠CDB

3. a) square
   b) rhombus
   c) rectangle

4. a) True
   b) True
   c) True

5. a, b and d are parallelograms, but c is not, since it is not a quadrilateral and its sides are not parallel.

6. a, b and d have point symmetry.

7. a) 2 cm
   b) Trapezoid

8. a) AB = CD
   b) No
   c) 110°
   d) 100°

9. a) Yes
   b) 2 ft.
   c) 90°

10. A

GEOMETRY: *Answers to* Chapter 8

1. a) True
   b) True
   c) False
   d) True
   e) False

2. a) Triangle AFG
   b) Line Segment CA

3. a) intersecting
   b) parallel
   c) the other
   d) translation

4. a) translation
   b) glide reflection
   c) glide reflection
   d) translation
   e) glide reflection

5. a) reflection image
   b) rotation image
   c) translation image

6. a) 60°
   b) point A

7. Yes, n = 6

8. a) Yes
   b) Yes, but not through the line shown.

9. a) No
   b) Yes

10. a) Yes
    b) No

GEOMETRY: *Answers to* Chapter 8

1. a) False
   b) True
   c) True
   d) False
   e) False

2. a) Reflection
   b) Rotation

3. a) Rotation
   b) Glide reflection
   c) Congruent

4. a) Translation
   b) Reflection
   c) Glide reflection
   d) Rotation
   e) Rotation

5. a) Reflection (line)
   b) Rotation
   c) Translation

6. a) 120 degrees
   b) The center of the circle.

7. Yes, $n = 3$

8. a) Yes
   b) No, when reflected about the line, the bottom of the figure does not coincide with the top of the figure.

9. The interstection of its diagonals.

10. The diagonal of a square forms two congruent triangles, hence the triangles coincide.

GEOMETRY: *Answers to* Chapter 9

1. a) False
   b) True
   c) True
   d) True
   e) False

2. 21 sq. cm

3. a) △GHE
   b) EFGH
   c) EFGH

4. a) 2 units
   b) $\sqrt{6}$ units
   c) 4 square units
   d) 6 square units
   e) 10 square units
   f) 3.16 units

5. $a^2 + 2ab + b^2$

6. a) 4 ft^2
   b) 576 inch^2.

7. a) 144 inches
   b) 12 feet

8. 12 cm^2

9. a) 5 cm
   b) 34 cm^2
   c) 59 cm^2

10. a) The area of a right triangle is half the product of its legs.
    b) The area of a polygonal region is equal to the sum of the areas of its nonoverlapping parts (the Area Postulate).
    c) Substitution.
    d) Substitution.

GEOMETRY: *Answers to* Chapter 9

1. a) True
   b) True
   c) False
   d) True
   e) False

2. a) 12 sq. cm
   b) 8 sq. cm
   c) 2 cm

3. a) GJH
   b) EGH

4. a) 2 cm
   b) 1 cm
   c) 6 cm^2
   d) 15 cm^2

5. $a^2 + 2ab + b^2$

6. a) a(b)
   b) a(b+ c)
   c) 10 sq meters

7. a) 8 cm
   b) 48 sq. cm

8. c, the trapezoid has the greatest area

9. Areas

10. a) The area of a right triangle is half the product of its legs.
    b) The area of a polygonal region is equal to the sum of the areas of its nonoverlapping parts (the Area Postulate)
    c) Subtraction
    d) Substitution
    e) Substitution

GEOMETRY: *Answers to* Chapter 10

1. a) True
   b) True
   c) False
   d) True
   e) True

2. $\dfrac{a}{b} = \dfrac{c}{d}$ (Given)

   $bd\left(\dfrac{a}{b}\right) = bd\left(\dfrac{c}{d}\right)$ (Multiplication)

   $\dfrac{abd}{b} = \dfrac{bcd}{d}$ so $ad = bc$ (Substitution)

3. a) a, b, c, d
   b) b, c
   c) a, d

4. a) Yes
   b) Yes
   c) 36 units
   d) 48 units

5. In $\triangle ABC$, $AB = AC$ and $DE \parallel BC$. (Given)

   $\dfrac{AD}{AB} = \dfrac{AE}{AC}$

   (If a line parallel to one side of a triangle intersects the other two sides in different points, it cuts off segments proportional to the sides.)

   $\dfrac{AD}{AB} = \dfrac{AE}{AB}$
   (Substitution)

   $AD = AE$
   (Multiplication)

6. $\angle A = \angle E$ and $\angle B = \angle B$. (If two angles of one triangle are equal to two angles of another triangle, the triangles are similar).

7. a) 20 sq units
   b) 180 sq. units
   c) 16 units

8. Two triangles similar to a third triangle are similar to each other.

9. a) The ratio of the perimeters of two similar polygons is equal to the ratio of the corresponding sides.
   b) 6 units
   c) 12 units
   d) 4 units

10. No, the sides are not in the same ratio.

GEOMETRY: *Answers to* Chapter 10

1. a) False
   b) True
   c) True
   d) False
   e) False

2. a) $\dfrac{4}{x} = \dfrac{x}{16}$ or $\dfrac{4}{8} = \dfrac{x}{24}$
   b) x = 8.

3. a) $\dfrac{4}{x} = \dfrac{8}{24}$
   b) 12

4. ∠B = ∠B' and ∠C = ∠C'. (If two angles of one triangle are equal to two angles of another triangle, the triangles are similar).

5. Two triangles similar to a third triangle are similar to each other.

6. a) $\dfrac{BG}{EH}$
   b) $\dfrac{BG}{EH} = \dfrac{AB}{DE} = \dfrac{BC}{EF} = \dfrac{AC}{DF}$
   (Any one will do.)
   c) 4 units
   d) 20 units

7. a) The ratio of the perimeters of two similar polygons is equal to the ratio of the corresponding sides.
   b) 6 units
   c) 24 units
   d) 8 units

8. 2 meters

9. a) Given
   b) If an angle is bisected, it is divided into two equal angles.
   c) Given
   d) If two sides of a triangle are equal, the angles opposite them are equal.
   e) Substitution
   f) Equal alternate interior angles mean that lines are parallel.
   g) If a line parallel to one side of a triangle intersects the other two sides in different points, it divides them in the same ratio.

10. 6

GEOMETRY: *Answers to* Chapter 11

1. a) False
   b) True
   c) True
   d) True
   e) False

2. a) AC
   b) BO
   c) ABC ~ BOC ~ AOB
   d) $\dfrac{BO}{CO}$ or $\dfrac{AB}{CB}$

3. $c^2 = a^2 + a^2 = 2a^2$,
   so $c = \sqrt{2a^2} = \sqrt{2}a = a\sqrt{2}$

4. a) 64 square meters
   b) 4 meters
   c) $4\sqrt{3}$ meters

5. a) Opposite, Adjacent
   b) Opposite, Hypotenuse
   c) Adjacent, Hypotenuse

6. a) 0.996
   b) 0.695
   c) 0.259
   d) 0.268
   e) 0.866

7. a) $\angle A = 12.312°$
   b) $\angle A = 0.705°$
   c) $\angle B = 14.036°$
   d) $\angle C = 82.760°$
   e) $\angle D = 45.000°$

8. $\dfrac{y_2 - y_1}{x_2 - x_1}$

9. a) 60°
   b) 46°
   c) 74°

10. a) 1
    b) –1
    c) e
    d) d
    e) a and c
    f) b

GEOMETRY: *Answers to* Chapter 11

1. a) True
   b) False
   c) True
   d) False
   e) False

2. a) 2 units
   b) $\frac{5}{2}$
   c) 10.5 units
   d) 12.5 units

3. a) $2x^2$
   b) $x\sqrt{2}$ units
   c) $10\sqrt{2}$ meters

4. a) 100 square centimeters
   b) 5 centimeters
   c) $5\sqrt{3}$ centimeters

5. When x = y or when A = 45°.

6.

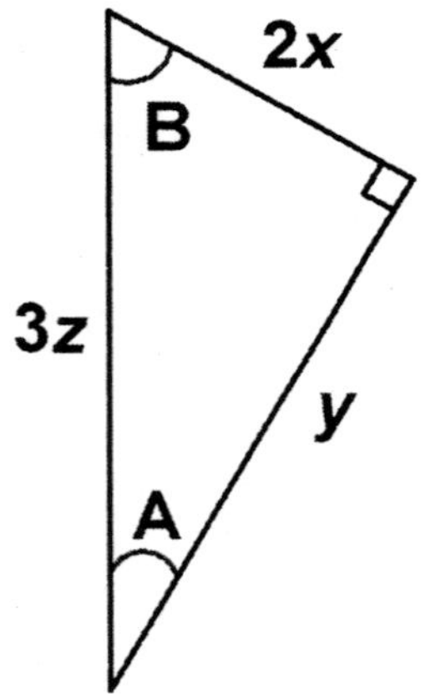

7. a) $\frac{6}{10}$ *or* $\frac{3}{5}$
   b) $\frac{8}{6}$ *or* $\frac{4}{3}$
   c) $\frac{6}{8}$ *or* $\frac{3}{4}$
   d) $\angle A = 37°$
   e) $\angle B = 53°$

8. 17°

9. 32.0 units

10. a) b
    b) a
    c) 0
    d) undefined
    e) $-\frac{1}{4}$

GEOMETRY: *Answers to* Chapter 12

1. a) False
   b) True
   c) True
   d) True
   e) False

2. a) Perpendicular lines form right angles.
   b) Two points determine a line.
   c) All radii of a circle are equal.
   d) Reflexive.
   e) HL.
   f) Corresponding parts of congruent triangles are equal.
   g) If a line divides a line segment into two equal parts, it bisects the line segment.

3. a) A line tangent to a circle intersects the circle in exactly one point.
   b) 90°
   c) If a line through the center of a circle is perpendicular to a chord, it also bisects it.

4. a) intercepted arc
   b) two points
   c) equal
   d) equal

5. a) AB = BC = CA
   b) arcAB = arcBC = arcCA or arcACB = arcACB = arcBAC
   c) Yes
   d) 120°
   e) 240°

6. *PROOF*
Vertical angles ∠AOB and ∠COD are central angles of both circles.
(Given)

∠AOB = ∠COD
(Vertical angles are equal.)
Measure of *arc*AB = ∠AOB and the measure of *arc*CD = ∠COD
(A minor arc is equal in measure to its central angle.)
Measure of *arc*AB = Measure of *arc*CD
(Substitution)

7. a) Tangent
   b) Chord
   c) Secant
   d) Secant

8. Construction of lines through P through to O.

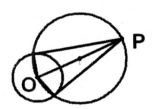

9. 3.31 cm

10. a) *arc*ED
    b) four
    c) 60°

12A-1

GEOMETRY: *Answers to* Chapter 12

1. a) True
   b) True
   c) False
   d) False

2. a) No
   b) Yes
   c) 65°
   d) No
   e) 32.5°

3. ABCD is a trapezoid and
   arcAB = arcCD.
   (Given)
   AB = CD
   (In a circle, equal minor arcs have equal chords.)
   ABCD is isosceles.
   (A trapezoid is isosceles if its legs are equal)
   ∠B = ∠C
   (The base angles of an isosceles trapezoid are equal.)

4. a) 100°
   b) 140°

5. a) Yes
   b) 226°
   c) 37°

6. a) 180°
   b) 90°
   c) AB = BC
   d) Yes

7. a) 40°
   b) arcAB = 45°, arcAC = 135°, arcBD = 135° and arcCD = 45°.
   c) 45°, 45°, 135° and 135°

8. a) All radii of a circle equal.
   b) The tangent segments to a circle from an external point are equal.
   c) If a line is tangent to a circle, it is perpendicular to the radius drawn to the point of contact.

9. a) Half of a circle.
   b) Less than half of a circle.
   c) More than half of a circle.

10. Measure of arcABC = 240°.

12B-1

GEOMETRY: *Answers to* Chapter 13

1. a) True
   b) False
   c) False
   d) True

2. a) Yes
   b) Yes
   c) No
   d) Yes

3. Circumscribe a circle about the triangle.

4. A, the square.

5. a) Yes
   b) HE, HF and HG
   c) No, circle H is inscribed in $\triangle ABC$.
   d) Yes

6. a) AF, BG and CE
   b) Yes
   c) Point D
   d) Yes

10. Angle Bisectors

7. a) Yes
   b) Point D
   c) 90°
   d) Because $\angle AED$ and $\angle DGA$ are right angles. (A quadrilateral is cyclic iff a pair of its opposite angles are supplementary.)

8. $\dfrac{AX}{XB} \cdot \dfrac{BY}{YC} \cdot \dfrac{CZ}{ZA} = 1$

9. $x = 12$

GEOMETRY: *Answers to* Chapter 13

1. a) True
   b) True
   c) True
   d) True

2. A and B are cyclic.

3. A copied triangle, with a circle inscribed inside

4. a) AF, BD, CE
   b) HM, IK, JL
   c) N
   d) G

5. $\dfrac{AX}{XB} \bullet \dfrac{BY}{YC} \bullet \dfrac{CZ}{ZA} = 1$

6. a) 3
   b) $\dfrac{1}{2}$
   c) $\dfrac{2}{3}$
   d) 4
   e) 6

7. An equilateral triangle.

8. a) True
   b) True
   c) False
   d) False

9.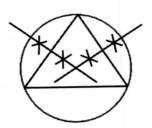

10. Since $\left(\dfrac{2Y}{3X}\right)\left(\dfrac{15X}{Y}\right)\left(\dfrac{2}{5}\right) = 4$, not 1, the cevians are not concurrent.

GEOMETRY: *Answers to* Chapter 14

1. a) False
   b) True
   c) False
   d) True

2. a) Zero
   b) Regular Pentagon
   c) 10 cm
   d) 72°

3. 5 inches

4. a) 58.78 meters
   b) 60 meters
   c) 587.8 meters
   d) 600 meters

5. a) 237.76 square meters
   b) 25,980.76 square meters or
   $15,000\sqrt{3}$ square meters

6. a) $n\sin\dfrac{180}{n}$
   b) $n\sin\dfrac{180}{n}\cos\dfrac{180}{n}$
   c) limit
   d) $2\pi r$

7. a) 2π meters
   b) 4π meters
   c) $x\pi$ meters
   d) $\dfrac{y\pi}{2}$ meters

8. a) limit
   b) πr^2
   c) radii

9. a) 8 square units
   b) 4π square units
   c) $4\pi - 8$ square units

10. 300π square inches

GEOMETRY: *Answers to* Chapter 14

1. a) True
    b) False
    c) True
    d) False

2. a) One
    b) Regular Hexagon
    c) 18 cm
    d) 60°

3. 10 inches

4. a) 30 meters
    b) 29.39 meters
    c) 293.9 meters
    d) 300 meters

5. a) 64.95 square meters
    b) 5,944.10 square meters

6. a) N
    b) M
    c) circumference
    d) circumference

7. a) 36π cm^2
    b) π cm^2
    c) 20π cm^2

8. a) Area
    b) Area
    c) Sector

9. a) 32 square units
    b) 16π square units
    c) 2π units

10. $\dfrac{100\pi}{3}$ square inches

GEOMETRY: *Answers to* Chapter 15

1. a) False
   b) True
   c) True
   d) True

2. a) 185 units
   b) 697 units
   c) 689.2 units
   d) 680 units

3. a) Rectangles
   b) Right Triangles
   c) 9 units
   d) 108 cubic units

4. 297.21 cubic centimeters

5. a) 300π cubic inches
   b) 1000π cubic inches

6. a) eight
   b) four
   c) r^3
   d) sphere

7. a) 64π m^2
   b) 201 m^2
   c) $\dfrac{64}{3}\pi$ cubic meters
   d) 67.02 cubic meters

8. a) Not always similar (student sketch of two non-similar cones).
   b) Always similar.
   c) Not always similar (student sketch of two non-similar square pyramids).
   d) Always similar.

9. The cube.

10. a) 3^2 or 9
    b) 3^3 or 27

GEOMETRY: *Answers to* Chapter 15

1. a) True
   b) True
   c) False
   d) False

2. a) $\sqrt{x^2 + y^2}$
   b) $\sqrt{x^2 + y^2 + z^2}$
   c) $\sqrt{y^2 + z^2}$
   d) $\sqrt{x^2 + z^2}$

3. a) Rectangles
   b) Trapezoids
   c) 9 units
   d) 288 cubic units

4. 528.80 cubic centimeters

5. a) 250π cubic inches
   b) 16π cubic inches

6. a) octahedron
   b) tetrahedron
   c) volumes
   d) point

7. a) 36π square meters
   b) 113.10 square meters
   c) 36π cubic meters
   d) 113.10 cubic meters

8. a) Not always similar (student sketch of two non-similar cones).
   b) Always similar.
   c) Not always similar (student sketch of two non-similar square pyramids).
   d) Always similar.

9. The icosahedron.

10. a) 5^2 or 25
    b) 5^3 or 125

GEOMETRY: *Answers to* Chapter 16

1. a) True
   b) False
   c) True
   d) True

2. a) Both
   b) Both
   c) Both
   d) Euclidean

3. a) 186°
   b) The sum of the two acute angles is greater than 90°.
   c) 90°
   d) 36° + 36° = 72°
   e) 36°
   f) No.
   g) 192°

4. a) One
   b) A line is a great circle and there is only one great circle on the sphere.
   c) Lines are great circles and great circles divide a the sphere into two equal hemispheres. So any two great circles must intersect, therefore are not parallel.

5. a) Two points determine a line.
   b) Definition of Saccheri quadrilateral.
   c) Definition of Saccheri quadrilateral. (Both are right angles.)
   d) Reflexive.
   e) △BAD ≅ △ABC (SAS)
   f) Corresponding sides of congruent triangles
   g) Corresponding angles of congruent triangles
   h) Subtraction of angles
   i) △BCD ≅ △ADC (SAS)
   j) Corresponding angles of congruent triangles

6. a) Right
   b) Acute
   c) Obtuse

7. a) A Saccheri quadrilateral.
   b) Acute summit angles.
   c) The line segment joining the midpoints of the base and summit of a Saccheri quadrilateral is perpendicular to both of them.

8. a) True
   b) False

9. In Lobachevskian geometry, the sum of the angles of a triangle are less than 180°. If one angle of a triangle is 90° then the sum of the other two angles must be less than 180° - 90° = 90°. Complementary angles sum to 90°, not less than 90°.

10. A *birectangular quadrilateral* is a quadrilateral that has two sides perpendicular to a third side.

GEOMETRY: *Answers to* Chapter 16

1. a) False
   b) True
   c) True
   d) False

2. a) 4 times greater
   b) $\dfrac{1}{4}$
   c) 8π units
   d) 6π units
   e) 64π square units
   f) 8π square units

3. a) 186°
   b) Sphere Geometry or Riemann
   c) 60°
   d) 36° + 36° + 36° = 108°
   e) 90°
   f) Yes
   g) No

4. a) An infinite number.
   b) Yes
   c) No, all great circles intersect.
   d) Two hemispheres.

5. a) The Ruler Postulate.
   b) Two points determine a line.
   c) It is a birectangular quadrilateral whose legs are equal.
   d) The summit angles of a Saccheri quadrilateral are equal.
   e) The "whole is greater than the part" theorem.
   f) Substitution.
   g) An exterior angle of a triangle is greater than either remote interior angle.
   h) The transitive property.

6. a) Exactly one
   b) More than one
   c) No

7. a) Obtuse
   b) AB
   c) Yes

8. a) False
   b) True

9. If two angles of two triangles are equal, the third angles are not necessarily equal since the sum of the angles of a triangle are less than 180°. The sum of the angles of each of the two triangles may not be the same.

10. *Polar Points* are the two points of intersection of a sphere with a line through its center.

GEOMETRY: *Answers to* Mid-Term

1. a) True
   b) False
   c) False
   d) True

2. If she speaks Spanish and English, then she is bilingual.

3. a) Subtraction Property
   b) Division Property
   c) Substitution Property
   d) Addition Property

4. △ABC is equilateral
   (Given)
   AB = BC
   (Def. of equilateral triangle)
   BD bisects ∠ABC
   (Given)
   ∠CBD = ∠ABD
   (Def. of Angle Bisected)
   BD = BD
   (Reflexive)
   △ABD ≅ △CBD
   (SAS)

5. a) (1,2).
   b) (4,2)
   c) 3 units
   d) 4 units
   e) 5 units

6. ∠B

7. a) AC
   b) 180°
   c) 180°
   d) ∠2, ∠8
   e) ∠2, ∠9

8. a) Copy line *j*.
   b) Find the midpoint M of *j*.
   c) Construct a line perpendicular to *j*, through point M.

9. a) square
   b) rhombus
   c) rectangle
   d) volumes

10. a) translation
    b) glide reflection
    c) glide reflection
    d) translation
    e) glide reflection

GEOMETRY: *Answers to* Mid-Term

1. a) approximately 24°
    b) approximately 101°
    c) approximately 56°
    d) No
    e) No

2. a) If you are 18
    b) then you can vote
    c) "you are 18", is a. "then you can vote", is b.
    d) If you can vote, then you are 18.
    e) Yes

3. a) No
    b) Yes
    c) Yes
    d) 95°
    e) ∠CGD + ∠DGE = ∠CGE
    f) Yes, if a ray bisects an angle, the angle is divided into two equal angles.
    g) Yes

4. AC bisects ∠DAB and ∠BCD
    (Given)
    ∠DAC = ∠BAC
    (Def. of Angle Bisected)
    AC = AC
    (Reflexive)
    ∠DCA = ∠BCD
    (Def. of Bisected)
    △ABC ≅ △ADC
    (ASA)

5. Sides and Angles

6. a) ∠2, ∠4, ∠7, ∠8, ∠9, ∠11
    b) ∠1, since the side opposite is longer.
    c) No

7. a) ∠1 and ∠5, ∠2 and ∠6, ∠3 and ∠7, ∠4 and ∠8
    b) ∠2 and ∠7, ∠4 and ∠5
    c) ∠2 and ∠5, ∠4 and ∠7

8. a) Copy the line *l* and point P above
    b) Construct a line parallel to *l*, through point P.

9. a) Rhombus
    b) Square
    c) Rectangle

10. a) Rotation
    b) Glide reflection
    c) Congruent

GEOMETRY: *Answers to* Final

1. a) True
   b) True
   c) True
   d) True

2. 24 square centimeters

3. a) No
   b) No (not as labeled)
   c) No
   d) Yes

4. a) ∠3, ∠8, ∠6
   b) ∠2, ∠4, ∠5, ∠7

5. a) Corresponding angles are equal when parallel lines are cut by a transversal.
   b) AA similarity.
   c) 6 units.
   d) 2:3 or $\frac{2}{3}$.

6. Circumscribe a circle about the triangle.

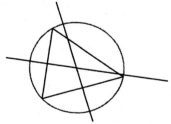

7. a) 58.78 meters
   b) 60 meters
   c) 587.8 meters
   d) 600 meters

8. a) Yes
   b) Yes
   c) Eight
   d) 10 inches

9. a) 324π square meters
   b) 1017.88 square meters
   c) 972π cubic meters
   d) 3,053.63 cubic meters

10. a) True
    b) False
    c) False
    d) True

GEOMETRY: *Answers to* Final

1. a) False
   b) True
   c) True
   d) False

2. In △ABC, AB = AC and DE // BC.
   (Given)

   $$\frac{AD}{AB} = \frac{AE}{AC}$$

   (If a line parallel to one side of a triangle intersects the other two sides in different points, it cuts off segments proportional to the sides.)

   $$\frac{AD}{AB} = \frac{AE}{AB}$$
   (Substitution)

   AD = AE
   (Multiplication)

3. a) Corresponding angles.
   b) AA similarity.
   c) 6 units
   d) 6 units
   e) 9:4 or $\frac{9}{4}$.

4. a) For example:

   b) Depend on answer to a).
   c) A rotation of 180°.

5. A copied triangle, with a circle inscribed inside

   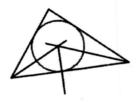

6. a) 180°
   b) 90°
   c) AB = BC
   d) 45°.

7. a) 30 meters
   b) 29.39 meters
   c) 293.9 meters
   d) 300 meters

8. a) 36π square meters
   b) 113.10 square meters
   c) 36π cubic meters
   d) 113.10 cubic meters

9. a) Rectangles
   b) Trapezoids
   c) 9 units
   d) 288 cubic units

10. a) A birectangular quadrilateral is a quadrilateral that has two sides perpendicular to a third side. A saccheri quadrilateral is a birectangular quadrilateral whose legs are equal.
    b) Greater circles are the "lines" of sphere geometry. Since every great circle intersects all other great circles of the sphere, there are no parallel lines.